OUTRAM'S RIFLES

Lieut.-General Sir James Outram, G.C.B.
(*Courtesy of the National Portrait Gallery*)

[*frontispiece*

OUTRAM'S RIFLES

A History of the 4th Battalion,
6th Rajputana Rifles

H. G. RAWLINSON, C.I.E.

The Naval & Military Press Ltd

Published by
The Naval & Military Press Ltd

CONTENTS

APPENDIXES:

ILLUSTRATIONS

INTRODUCTION

THE History of Outram's Rifles, which I have undertaken to write at the request of the Officers of the Regiment, is a companion volume to the history of their sister battalion, Napier's Rifles, published a few years ago. Both battalions have had a long and distinguished record, going back to the beginnings of British rule in Western India. Absolutely loyal during the Mutiny, they have shed their blood for the Raj in many distant lands, and their magnificent services during the Great War won the thanks of the General Officers under whom they fought. At the present juncture, when old historic landmarks are fast disappearing, together, alas, with many of those who participated in these stirring events, it is only fitting that so fine a story should be rescued from oblivion.

In writing this history, my chief authorities have, of course, been the regimental records, and the earlier compilation of Captain W. A. M. Wilsón (going down to 1893), who also wrote an admirable account of the regiment's operations in the Burma Campaign. Captain Wilson died of enteric fever at Deolali in 1895, shortly after the publication of the latter work. He was D.A.A.G. for Musketry and Commandant of the School of Musketry at the time of his death, by which the Indian Army suffered a great loss. Of both these I have freely availed myself. Other authorities will be found quoted in the footnotes to each chapter. For the Battle of Kirkee, I have made use of some of the material collected for *Napier's Rifles*. For the Palestine Campaign, the Official

History by Captain Cyril Falls has been of the greatest value, but I have chiefly depended upon the diaries and records supplied by the officers of the battalion who took part. I have especially to thank General Sir Robert Scallon, G.C.B., Brigadier-General G. R. Cassels, C.B., D.S.O., Lieut.-Colonel B. G. B. Kidd, D.S.O., and Lieut.-Colonel R. P. T. Ffrench, M.C., for their invaluable assistance. The Controller of His Majesty's Stationery Office has kindly permitted the reproduction, from the *Official History of the War in Egypt and Palestine*, of the map to illustrate the action at Nebi Samwil; Messrs. Constable & Co. have permitted me to use the picture of Nebi Samwil from Mr. W. T. Massey's *How Jerusalem was won*; and Miss M. F. Outram has allowed me to reproduce an early portrait of her grandfather, which has already appeared in her *Margaret Outram*, published by John Murray. The portrait of Sir James Outram which forms the frontispiece is from the original in the National Portrait Gallery.

H. G. RAWLINSON

THE BATTLE OF KIRKEE AND THE BIRTH OF THE REGIMENT

CHAPTER I

(1817-8)

THE BATTLE OF KIRKEE AND THE BIRTH OF THE REGIMENT

IN order to understand the circumstances which led to the battle of Kirkee, it will be necessary to glance briefly at the history of the Marathas. In the earlier part of the eighteenth century, the Peshwas, the Brahmin Prime Ministers of Poona, had succeeded in usurping the power of the Maratha kings, and had reduced them to the position of *rois fainéants*. But after the defeat of the Marathas at the battle of Panipat in 1761, their influence began to decline, and by the end of the century Poona was in a state of wild confusion and anarchy. In 1802, the last Peshwa, Bajirao II, was deposed by his rivals. He fled to Bassein, where he formed a treaty with the British by which he bartered his liberty as the price of his restoration. He agreed to receive a subsidiary force at Poona, pay an annual tribute, and recognize the suzerainty of the Company as exercised through the Resident. Ten years later he further agreed to raise a body of Indian Infantry, to be known as the Poona Auxiliary Force, commanded and drilled by Europeans, and paid direct from the Poona Treasury. This force was put under the command of Captain John Ford, originally an officer of the Madras establishment and afterwards for some years on the Resident's staff at Poona, assisted by a number of picked British officers. His men, excepting a small leavening of Marathas, were chiefly raised in the Company's provinces in Hindustan, and on joining the brigade swore fidelity to the Peshwa, 'so long as he kept in alliance with the British'. The force was stationed in Dapuri, about six miles north-west of Poona. Here Ford and his officers lived in great state, in a fine castellated building which has since served as Government House, a brewery, and a P.W.D. store. His Brigade Major was Captain (afterwards General) Peter Lodwick, the discoverer of the hill-station of Mahableshwar. Ford was on excellent terms with the Maratha generals and sirdars, including Moro Dixit, and the gallant Bapu Gokhle, who commanded the Peshwa's forces. But Mountstuart Elphinstone, the distinguished scholar, diplomatist and soldier, who had succeeded Colonel Barry Close as British Resident at Poona in 1810, was not prepared to trust Bajirao's fair words and protestations. A deliberate affront had been offered to the English when Gangadhar Shastri, the Gaikwar's minister, who was travelling under a safe-conduct given by the British Government, was insolently murdered in the streets of Pandharpur by the hired assassins of Trimbakji Dengalia, the Peshwa's low favourite. Under great pressure, Trimbakji was given up and imprisoned in Thana Fort, but he had, with the Peshwa's connivance, escaped, and was now in hiding in the vicinity of Poona. Relations were becoming more and more strained every day, and war was almost inevitable. The available troops, in the event of a rupture, were less than three thousand in all, distributed as follows :—

In the Garpir cantonment, where the Imperial Bank and Post Office now stand, were the 2nd Battalion 1st Bombay Native Infantry, 2nd Battalion 6th Bombay Native Infantry, and the 1st Battalion 7th Bombay Native Infantry, with four guns, under the command of the veteran Lieut.-Colonel Burr ; at the Sangam Residency, now the District Judge's bungalow, was the Resident's escort, consisting of two companies of Bengal Native Infantry ; at Dapuri, four miles distant, was Ford's Brigade with three guns. Thirty-six miles from Poona, on the Ahmednagar road, was the Sirur brigade in cantonments at that place. The Peshwa was now becoming every day more insolent in his demeanour, troops were pouring into Poona, and it became unsafe for anyone belonging to the British camp to quit their lines. Lieutenant Shaw, of the Auxiliaries, was, on November 1st, dismounted, within two miles of the Residency, by three of the Peshwa's horsemen, and speared in the thigh. On the same day, the British brigade at Poona changed its ground from the immediate vicinity of the city to a more tenable position at a distance of about three miles, leaving a small guard in charge of their cantonments. This measure had become absolutely necessary, not only on account of the weakness of a position towards which the enemy was closing in from all directions, thus depriving the brigade of the advantage of its discipline, but likewise owing to the unremitting endeavours of the Peshwa to corrupt the fidelity of the sepoys, to which they were the more exposed in proportion to their vicinity to the city. The left of the new position rested on the Kirkee bridge over the Mulla river, and its right on the village of Kirkee. This position was rather too extensive for the troops which occupied it; but it had the advantage of commanding the water, and communicating directly with the Residency, and the great roads to Sirur and Bombay.

It happened, fortunately, that the Bombay European Regiment (the famous 'Bombay Toughs', afterwards the 2nd Battalion Dublin Fusiliers), with detachments of His Majesty's 65th Foot and of the Bombay Artillery, which were on their march from the coast to the Poona Division, had not yet passed that capital towards their northern destination. This reinforcement entered the camp at Kirkee on the 30th October, after a forced march of seventy-two miles from the bottom of the Ghats, with only a single three hours' halt *en route*. They entered the British lines travel-stained, but with ranks closed, band playing and colours flying. Not a man had fallen out, for all knew 'there must be a battle soon'.

As soon as the news of their arrival reached the Peshwa, he determined to throw off the mask. His preparations began about seven o'clock on the morning of the 5th; but in the early part of the day he sent out several messages calculated to lull the Resident's suspicions; such as, that his troops were alarmed by hearing that those at Kirkee were under arms; that he was about to perform a religious ceremony at the temple of Parbati, and that the troops were drawn out in honour of the occasion, to form a street as he passed. In the afternoon, when all was in readiness, the whole of his principal officers having assembled at his place, Vithuji Gaikwar, a personal servant of the Peshwa, was despatched to Elphinstone, by Gokhle's advice, to inform him that the assembly of troops at Poona was very offensive to the Peshwa; to desire him to send away the European regiment; to reduce the native brigade to its usual strength, after which it must occupy a position which the Peshwa would point out; and that if these demands were not complied with, he would withdraw from Poona and never return. Elphinstone denied the Peshwa's right to require the removal of the European regiment, and recommended that the Peshwa should send his troops to the frontier as he had promised, in which case all cause of complaint would be removed. A good deal more passed, as the conversation on the part of the messenger was intended to engage as much attention as possible; but he at last withdrew, warning the Resident of the bad consequences of his refusal. In the meantime the Peshwa's officers at the palace were despatched to their troops; Bajirao in person proceeded to Parbati Temple; and Vithuji Gaikwar had scarcely quitted the Residency when intelligence was brought that the army was moving out on the west side of the city. There was a momentary consultation about defending the Residency, but it was instantly abandoned as impracticable, and it was determined to retire to Kirkee, for which purpose the nature of the ground afforded great facility. The river Mutha, betwixt the Sangam and the village of Kirkee, forms two curves like the letter S inverted. The Residency and the village were both on the same side of the river, but at the former there was a ford, and near the latter a bridge; so that the party, by crossing at the ford, had the river between them and the Peshwa's troops the greater part of the way. The Peshwa's troops, now entering the gardens of the Residency, set fire to everything combustible and in a short time destroyed the buildings, furniture and records. From the Residency no part of the Maratha army was visible excepting bodies of infantry, which were assembling along the tops of the adjoining heights, with the intention of cutting off the Residency from the camp, and, having this object in view, they did not molest individuals. On ascending one of the eminences on which they were forming, the plain beneath presented at that moment a most imposing spectacle. This plain, then covered with grain, is terminated on the west by a range of small hills, while on the east it is bounded by the city of Poona and the small hills already partially occupied by the infantry. A mass of cavalry covered nearly the whole extent of it, and towards the city endless streams of horsemen were pouring from every avenue.

'Those only who have witnessed the Bore in the Gulf of Cambay,' says Grant Duff, the historian of the Marathas, 'and have seen in perfection the approach of that roaring tide, can form the exact idea presented to the author at the sight of the Peshwa's army. It was towards the afternoon of a very sultry day; there was a dead calm, and no sound was heard except the rushing, the trampling and neighing of the horses, and the rumbling of the gun-wheels. The effect was heightened by seeing the peaceful peasantry flying from their work in the fields, the bullocks breaking from their yokes, the wild antelopes, startled from sleep, bounding off, and then turning for a moment to gaze on this tremendous inundation, which swept all before it, levelled the hedges and standing corn, and completely overwhelmed every ordinary barrier as it moved.'

At a distance of about three miles to the north-westward of the city of Poona lie some hills, between which passes the road to Bombay. In that direction the Peshwa's army was assembled since morning, in expectation of orders for attacking the British force, to which it was opposite.

A high ridge of ground extended between the two positions, which were distant more than two miles. On each side flowed, in the lower ground, the Mutha river, which doubled round the rear of the British camp. Many ravines and water-courses joined the bed of the river from the high ground round which it flowed, and were calculated to impede any distant flank movement along the bank.

The enemy's position was on strong, commanding ground. In front were a rivulet and some walled gardens; his left was at the Ganeshkhind hills, his right on the Residency, and in his rear a chain of mountains. The Vinchur horse were on the left, the guns and infantry in the centre, and large bodies of cavalry on the right and along the rear. The Maratha army was immediately commanded by Gokhle, the Peshwa having retired to the top of Parbati, to have a distant and elevated view of the action. Moro Dixit, the Minister, held a respectable command, though more considered as a politician than a soldier.

The veteran Lieut.-Colonel Burr commanded the British force, and had been engaged since his removal to Kirkee in the establishment of a post for the security of his stores, treasures, provisions, and sick, before the expected crisis. He now found himself enabled to leave them under the protection of the 2nd Battalion of the 6th Bombay Native Infantry, greatly weakened by small detachments, but reinforced by such details as he considered unfit to bring into action. This position, in which were placed two iron twelve-pounders, was committed to the charge of Major Roome.

During the retreat of the Resident from the Sangam, Lieut.-Colonel Burr was required to move forward to meet the enemy, already in motion to commence the action. The Colonel was informed that Ford's battalion was marching to join him from its cantonment at Dapuri, situate about two miles distant in a westerly direction.

Elphinstone had personally reconnoitred the ground in front of the village of Kirkee and ascertained that there was a ford between that village and Dapuri which, although difficult, was practicable for six-pounders, three of which, manned by native artillerymen, belonged to the auxiliary force and were attached to Captain Ford's corps. It had been arranged, in the case of an attack, that Captain Ford was to join the brigade under Lieut.-Colonel Burr; and Elphinstone had been at pains to explain to all concerned the advantage of always acting on the offensive against Marathas.

The force accordingly selected a position about a mile further in advance. It was joined by the Resident with the guard from the Sangam, and formed, in expectation of receiving here the reinforcement under Captain Ford. In the centre were the Bombay European Regiment, the Resident's escort, and a detachment of the 2nd Battalion of the 6th Regiment;

on the right and the left were the 2nd Battalion of the 1st and the 1st Battalion of the 7th Regiments, and on each of their outer flanks two guns.

It was now about four p.m. ; and as the Dapuri Brigade approached, the force again advanced, while the enemy threw forward his cavalry in masses on the right and left, for the purpose of passing to the rear of the British force, between it and the river. Immediately afterwards, a brisk cannonade opened from their centre. The Dapuri Brigade was still about one thousand yards distant from the right of the line, and offered a tempting inducement to a corps of the enemy, who hoped to cut it off while yet detached. Moro Dixit commanded this body of cavalry, and is said to have been excited to undertake so brilliant an enterprise in consequence of some sarcastic remarks made by Gokhle, with reference to his former pursuits and his supposed disposition in favour of the English. As he approached the right flank of the Brigade, that wing was promptly thrown back, and when the distance was considered short enough for ample execution, an animated fire was commenced from the battalion and its three field-pieces, which obliged the enemy to desist and continue his movement towards Kirkee. He was here received by the two twelve-pounders which had been placed in position ; and, having lost the leader, Moro Dixit, who was shot in the mouth during the attack, this body of the enemy, discouraged from any further attempt in that quarter, turned towards the rear of the British line.

The Marathas, who had sent on their skirmishers, some of whom had already suffered from the fire of the light infantry, were surprised by this forward movement in troops, whom they had been encouraged to believe were already spiritless ; and a damp, which had been spreading over the whole army by the accidental breaking of the staff of the Jari Patka[1] before which they left the city, was now much increased. Gokhle, with the true spirit of a soldier, was riding from rank to rank, animating, encouraging and taunting as he thought most effectual ; but the Peshwa's heart failed him, and after the troops had advanced he sent a message to Gokhle desiring him

[1] The national standard of the Marathas.

'to be sure not to fire the first gun'. At the moment the British troops were halted, their guns were unlimbering—it was the pause of preparation and of anxiety on both sides; but Gokhle, observing the messenger from the Peshwa and suspecting the nature of his errand, instantly commenced the attack by opening a battery of nine guns, detaching a strong corps of rocket-camels to the right, and pushing forward his cavalry to the right and left. The British troops were soon nearly surrounded by horse; but the Maratha infantry, owing to this rapid advance, were left considerably in the rear, except a regular battalion under a Portuguese, named De Pinto, which had marched by a shorter route, concealed for a time under cover of the enclosures and was now forming with apparent steadiness, immediately in front of the 1st Battalion 7th Regiment, and the grenadiers of the 2nd Battalion 6th. No sooner, however, were their red coats and colours exposed to the view of the English sepoys than the latter, with one accord, pushed forward to close, and in their eagerness got detached from the rest of the line. Gokhle, hoping that they might either be disposed to come over or that he might be able to take advantage of their impetuosity, prepared a select body of six thousand horse, which, accompanied by the Jari Patka, and headed by several persons of distinction, had been held in reserve near his left, and were now ordered to charge. The Maratha guns ceased firing to let them pass; and they came down at speed, in a diagonal direction across the British front. Giving their fire, and receiving that of the line, they rode right at the 7th. Lieut.-Colonel Burr, took his post with the colours of that corps; it had long been his own battalion, he had 'formed and led' it for many years; he was then suffering under a severe and incurable malady, but he showed his wonted coolness and firmness in this moment of peril. He was the first to perceive the moving mass; he had just time to stop the pursuit of De Pinto's battalion, already routed, and to call to the men, who could not be dressed in line, to reserve their fire and prove themselves worthy of all his care. Fortunately, there was a deep slough, of which neither party were aware, immediately in front of the British left. The foremost of the horses rolled over, and many, before they could be pulled up, tumbled over

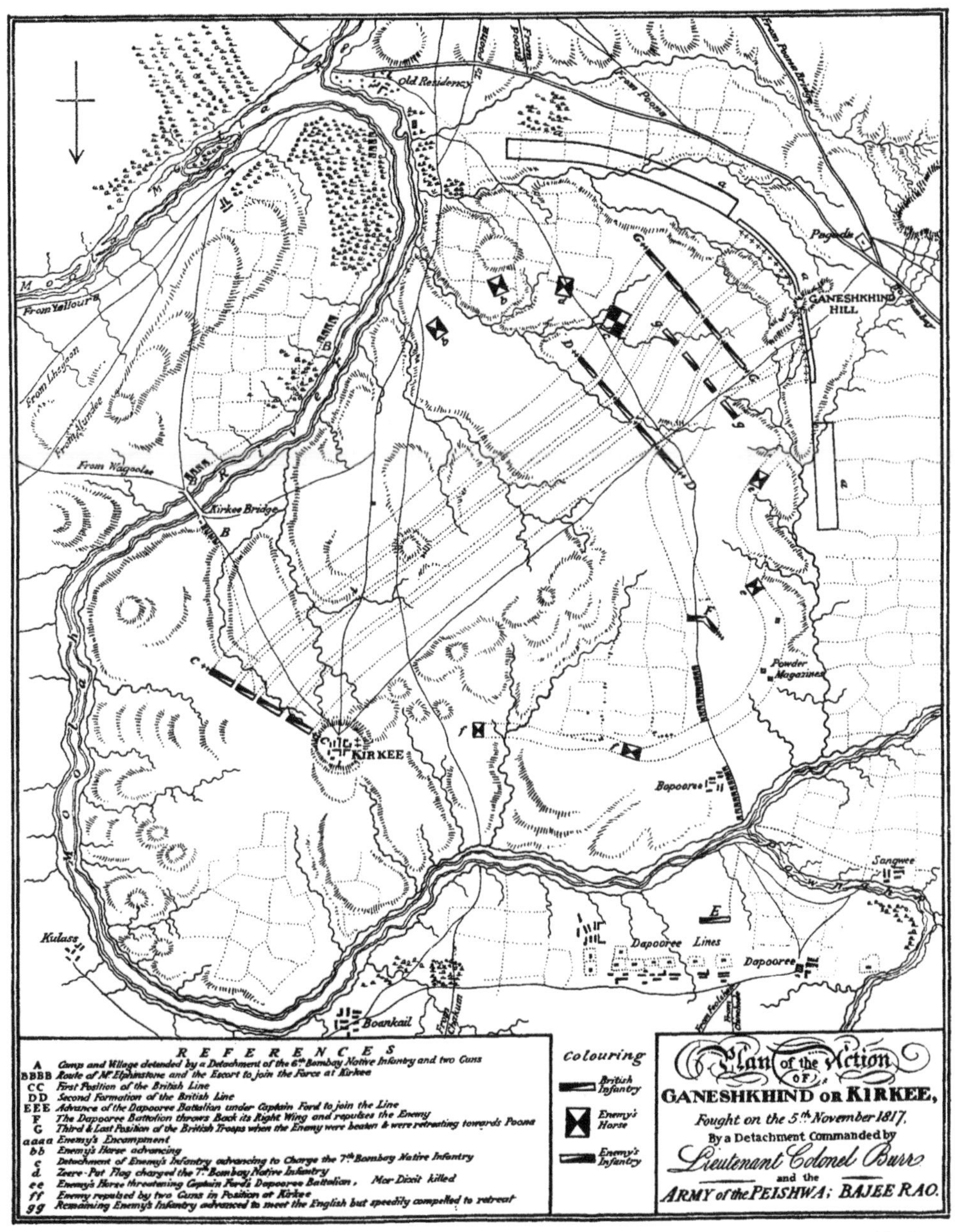
Plan of the Action
OF
GANESHKHIND OR KIRKEE,
Fought on the 5th November 1817,
By a Detachment Commanded by
Lieutenant Colonel Burr
and the
ARMY of the PEISHWA; BAJEE RAO.
REFERENCES
A Camp and Village defended by a Detachment of the 6th Bombay Native Infantry and two Guns
BBBB Route of Mr. Elphinstone and the Escort to join the Force at Kirkee
CC First Position of the British Line
DD Second Formation of the British Line
EEE Advance of the Dapooree Battalion under Captain Ford to join the Line
F The Dapooree Battalion throws Back its Right Wing and repulses the Enemy
G Third & Last Position of the British Troops when the Enemy were beaten & were retreating towards Poona
aaaa Enemy's Encampment
bb Enemy's Horse advancing
c Detachment of Enemy's Infantry advancing to Charge the 7th Bombay Native Infantry
d Zeere-Put Flag charged the 7th Bombay Native Infantry
ee Enemy's Horse threatening Captain Ford's Dapooree Battalion, Mor Dixit killed
ff Enemy repulsed by two Guns in Position at Kirkee
gg Remaining Enemy's Infantry advanced to meet the English but speedily compelled to retreat
Colouring
British Infantry
Enemy's Horse
Enemy's Infantry
Old Residency
From Poona
From Poonah Bridge
Pagoda
GANESHKHIND HILL
From Yellour
From Lhegaon
From Wagoolee
Kirkee Bridge
KIRKEE
Powder Magazines
Dapooree
Sangwee
Dapooree Lines
Kulass
Boankail
From Chakum

[face p. 10

those in front. The fire, hitherto reserved, was now given with great effect, numbers fell, the confusion became extreme, and the force of the charge was completely checked; a very small proportion came in contact with the bayonets, a few continued the attack in the rear, but many turned back; some galloped round the left as if to plunder the camp, but they were driven off by a few shots from the two iron guns at Kirkee, and the sepoys had nearly repulsed the attack before a company of Europeans could arrive to their support.

On the junction of Captain Ford's Brigade, which formed on the flank at which it arrived, little alteration was made in the previous dispositions. The additional guns from Dapuri being placed on the right flank, those which had formerly occupied that position were moved to the centre. At the same time, the light companies of the 7th Regiment, which had hitherto preceded the line, were sent to the rear to oppose the demonstrations of the enemy's horse, who had turned the right flank. Lieut.-Colonel Burr again advanced; and finding his line much galled by numerous skirmishers who occupied some garden enclosures, and a nullah in his front, he detached all the remaining light infantry to dislodge them; a service which was effected as it became dark.

The enemy had now resumed his original position and was engaged in drawing off his guns towards the city. The British force was in possession of the field of battle; and probably from prudence, as well as want of daylight, considered it inadvisable to persist in a further attack. The troops accordingly marched back to their camp at Kirkee, with the exception of the Dapuri corps, which returned to its isolated cantonments; so well satisfied were the victors that no molestation would be attempted during the night.

The Marathas in Captain Ford's Brigade deserted, and a part of the newly-raised auxiliary horse were, at their own desire, permitted to quit the British camp; but not one sepoy of the regular service left his colours. The number of the British troops engaged at the affair of Kirkee, including Captain Ford's battalion, was 2,800 rank and file, of whom about 800 were Europeans. Their loss was comparatively trifling, amounting only to 86 men in killed and wounded, 50 of whom were of the sepoys on the left. The Maratha army consisted of 18,000 horse and 8,000 foot, with 14 guns. They suffered considerably, having lost 500 men in killed and wounded; and though the proportion of horses killed on the spot was inconsiderable, a very great number were disabled. Amongst the sufferers was, as we have seen, the Minister Moro Dixit, who, by rather a strange fatality, had been mortally wounded by a grapeshot from one of the guns attached to the battalion of his friend, Captain Ford.

So ended the battle of Kirkee; not a great fight, perhaps, when compared with the bloody combats of Meeanee or Assaye, but one which proved decisively that the Indian sepoy, disciplined and led by English officers, was more than a match for the finest cavalry in India. An extract from a letter from Mountstuart Elphinstone to a friend, written on the battlefield, must close our account:

'Camp Kirkee,
November 11, 1817.

'The Peshwa, who had perhaps been flattered by Gokhle that all his preparations would be made without his getting into a scrape, now saw that he must throw off the mask. Accordingly, he sent a very bullying message to desire that I would move the cantonment to such place as he should direct, reduce the strength of the Native Brigade, and send away the Europeans; if I did not comply, peace could not last. I refused, but said I was most anxious for peace, and should not cross the river towards Poona, but if his army came towards ours we should attack it. Within an hour after, out they came with such readiness that we had only time to leave the Sangam with the clothes on our backs, and, crossing the river at a ford under Clieland's,[1] march off to the bridge with the river between us and the enemy, and a little firing but no real fighting. The Sangam, with all the records and all my books, journals, letters, manuscripts, etc., were soon in a blaze, but we got safe to the Kirkee bridge, and soon after joined the line. While the men and followers were fording, we went ourselves to observe the

[1] Clieland commanded the Residency Guard.

enemy. The sight was magnificent as the tide rolled out of Poona. Grant,[1] who saw it from the height over the powder cave, described it as resembling the Bore in the Gulf of Cambay. Everything was hushed except the trampling and neighing of horses, and the whole valley was filled with them like a river in flood. I had always told Colonel Burr that when war broke out we must recover our character by a forward movement that should encourage and fix our own men, while it checked our enemies, and I now, by a lucky mistake, instead of merely announcing that the Peshwa was at war, sent an order to move down at once and attack him. Without this, Colonel Burr has since told me, he certainly would not have advanced, However, he did advance, we joined, and, after some unavoidable delay, the Dapuri battalion joined too. When opposite to the nullah, where there used to be a plantain garden, we (injudiciously, I think) halted to cannonade, and at the same moment the enemy began from twelve or fifteen guns. Soon after his whole mass of cavalry came on at speed in the most splendid style. The rush of horse, the sound of the earth, the waving of flags, the brandishing of spears, were grand beyond description, but perfectly ineffectual. One great body, however, Gokhle and Moro Dixit and some others, formed on our left and rear, and when the 1st-7th was drawn off by its ardour to attack Major Pinto, who appeared on our left, and was quite separated from the European Regiment, this body charged with great vigour and broke through between it and the European regiment. At this time, the rest of the line was pretty well occupied with shot, matchlocks, and above all with rockets, and I own I thought there was a good chance of our losing the battle. The 1st-7th, however, though it had expended all its ammunition, survived the charge and was brought back to the line by Colonel Burr, who showed infinite coolness and courage ; and after some more firing, and some advancing, together with detaching a few companies to our right towards the little hill of Ganeshkind, we found ourselves alone in the field and the sun long set. I was at first for advancing to the water at the Sait's Garden, but was persuaded it was better to return to camp, which it was. If we

[1] Grant Duff, the historian of the Marathas.

had not made the move forward, the Peshwa's troops would have been quite bold, ours quite cowed, and we doubtful of their fidelity. We should have been cannonaded and rocketed in our camp, and the horse would have been careering within our pickets. As it is the Peshwa's army has been glad to get safe behind Poona, and we have been almost as quiet as if encamped on the Retee[1] at Delhi.'

As soon as General Smith, who commanded the Sirur brigade, found communications cut off, he advanced on Poona (November 13th). From the time his force quitted Sirur, he was followed by a flying party of Marathas, who, owing to his want of cavalry, harassed his march. He arrived on the evening of the 13th, and preparations were made to attack the Peshwa before daylight of the 15th. His army, having obtained a considerable addition by the junction of most of the southern jaghirdars, had come out a few days before and encamped with its left on the late cantonment of the British troops, and its right stretching along the Sholapur road for several miles. The intended attack, however, on the morning of the 15th, was postponed by General Smith in consequence of unforeseen difficulties at the ford. About sunset on the evening of the 16th an advanced brigade was ordered to cross the ford, and take up a position to the east of the Peshwa's army at the village of Ghorpuri, for the purpose of co-operating in an intended attack on the ensuing morning ; it was opposed by a body of the Peshwa's infantry, supported by parties of horse and two guns, but having succeeded in getting to its station, though with the loss of 84 men in killed and wounded, it was no longer molested during the night. In the morning, when General Smith moved towards the camp, he found it abandoned, and that the Peshwa had fled towards Satara. During the day Poona was surrendered.

The battle of Kirkee was in reality the ' baptism of fire ' for Outram's Rifles, for from the faithful sepoys of the Poona Auxiliary Force the regiment was originally constituted. Their conduct on the field had won them high praise. Colonel Burr on his despatch wrote : ' To Major Ford and the men of his fine

[1] Camping-ground.

brigade I feel the greatest obligations for the cheerfulness and anxiety they evinced to contribute to the general success of the day', and the Commander-in-Chief in a General Order stated that: 'The conduct of Captain Ford and the Brigade under his command is entitled to the Commander-in-Chief's cordial approbation.'

Lady Bartle Frere, writing at a time when the battle of Kirkee was still a living memory, speaks with admiration of the wonderful loyalty of the sepoys of Ford's Brigade. 'Old soldiers', she says, 'would tell how the fidelity of the sepoys resisted all the bribes and threats of Bajirao Peshwa, and thus, while the Peshwa hoped to effect his purpose by treachery, enabled Mr. Mountstuart Elphinstone to defer open hostilities —a matter of vital importance to the operations of Lord Hastings on the other side of India, in preparing for his great campaign against the Pindaris. The veterans would recount all the romantic incidents of the struggle which followed. . . . As they formed up in line of battle, they anxiously watched the native regiments coming up on their flank from Dapoorie, for that was the moment for successful treachery, if the native soldiers were untrue! Not a sepoy, however, in the British ranks wavered, though, before the junction was complete, a cloud of Mahratta cavalry dashed through the opening left between the two lines, enveloped either flank of the little army, and attacked the European regiment in the rear.'

Thanks to good discipline and fire-control, their casualties were trifling, one sepoy killed and seven wounded, and one British officer, Lieutenant Faulconer, severely wounded in the shoulder. Nine years later, their prowess on the field was acknowledged in an order of the Honorable the Governor in Council, issued from Head-quarters, Bombay Castle, on June 11th, 1827. It runs as follows:—

'The 23rd Regiment, Native Infantry (formerly 1st Battalion 12th Regiment, Native Infantry) having, with the exception of the European Officers, been originally formed from the late Poona 1st Auxiliary and Supernumerary Battalion of which about 500 were in action with the enemy at Kirkee on November 5th, 1817, the Honorable the Governor in Council is pleased in acknowledgment of the attachment of the Native Troops to the British interest on that trying occasion to authorize the honorary badge of " Kirkee " to be borne on the colours and appointments of the Regiment, as granted to the other corps distinguished on that occasion.'

Immediately after the battle they were reorganized under their old officers, Ford and Lodwick, as the Poona Auxiliary Infantry, and formed part of the force which was employed under General Pritzler in the early part of 1818 in reducing the Maratha strongholds, which are a characteristic feature of the Deccan. They were present at the capture of Lohogarh, Takuna, Tunga and Kuari: these formidable forts, almost impregnable to infantry attack, offered but a feeble resistance to the high-angle fire of the mortars and heavy guns brought against them. The expedition against Vasota, a fortress situated in the midst of wild country in a remote and unexplored region of the Western Ghats, was a more difficult undertaking. Here were confined two British prisoners of war, Cornets Wilson and Hunter, besides the wives and families of the Raja of Satara and his brothers, and it was said that the Peshwa had ordered the latter to be put to death rather than suffer them to fall into the hands of the British Government. The siege-train was dragged with infinite labour through thick jungles and deep valleys where natural barriers presented themselves at every hill, at which, as Grant Duff says, a handful of men might have arrested a host. At length the mortars opened, and the peal of the salvos reverberated from the surrounding rocks, but it was only after twenty hours' bombardment that the Commandant capitulated. The prisoners were rescued unharmed, and the two British officers presented a strange picture to their rescuers, with their long beards and coarse native clothing! The force now returned to Satara, where the Raja, Pratap Singh Bhosla, was reinstated with much pomp on the throne of his fathers under British suzerainty. The final scene in the drama was enacted on June 3rd, when the Peshwa surrendered to Sir John Malcolm, and was banished to Bithur. The Maratha

Empire had ceased to exist and the Deccan passed into British hands.

AUTHORITIES

Memoirs of the Operations of the British Army in India during the Mahratta War of 1817-19, by Lieut.-Colonel V. Blacker.

Grant Duff's *History of the Mahrattas.*

Colebrooke's *Life of Mountstuart Elphinstone.*

Unpublished papers in the Poona Daftar (Residency Records).

Regimental Records.

JAMES OUTRAM

CHAPTER II

(1818-1825)

JAMES OUTRAM

'In early manhood he reclaimed wild races by winning their hearts.'

AFTER the conquest of the Deccan, it was found necessary to reorganize the Bombay Army, in order to cope with the large accession of new territory, and amongst other measures, in 1820 it was decided to add a regiment of Native Infantry to the Infantry of the Line. It was to be denominated the 12th, and to consist of two battalions of 850 rank and file each, the first battalion to be stationed at Poona, and the second or Marine Battalion at the Presidency.[1] The old Poona Auxiliary Infantry was to be disbanded, and the new regiment was to be constituted partly from volunteers from the Poona Auxiliary Infantry, and partly by drafts from other regiments. The latter expedient, however, proved unnecessary. So strong was the *esprit de corps* among the sepoys that practically the whole of the 1st and Supernumerary battalions volunteered for the 1st Battalion 12th Native Infantry. On the disappearance of the Poona Auxiliary Force, the Commander-in-Chief issued the following General Order :—

'The Poona Auxiliary Infantry having been broken up by the late arrangements, the Honourable the Governor in Council has much satisfaction in expressing the approbation with which he has viewed the uniform good conduct of that valuable corps. The reports of the high state of discipline to which these troops have attained, and of their gallantry and fidelity during the war bear ample testimony to the zeal

[1] The Marine Battalion is now the 1st Battalion, 10th Baluch Regiment (formerly the 124th D.C.O. Baluchistan Infantry).

and ability of Major Ford, their Commanding Officer; Captain Lodwick, Major of Brigade; and Major Hickes, Captain Betts and Captain F. Hicks, and of the whole of the officers under their command.

'The Governor in Council requests these officers to accept of his thanks for their honourable exertions, and further requests that His Excellency the Commander-in-Chief will be pleased to communicate to the whole of the Native Officers and men, who lately composed the Corps of Poona Auxiliary Infantry, the high sense which the Governor in Council entertains of their past services, and his confidence that they will acquire fresh claims to the approbation of Government in the branches of the Army to which they have desired to transfer their services.'

It was ordered that the new battalion should wear dark green facings with silver lace, put on 'bastion fashion' (obliquely), and instead of a stock, they had the exclusive privilege of wearing a necklace of white beads (*Kantha*), from which they acquired their well-known nickname of the *Kanthawali-ki-Paltan*. The *kantha* was introduced by the high-caste Hindustani sepoys, and was universally worn in the old Bengal Army. The mutiny, however, effaced all feelings of pride in the badge, which had already been dropped in Bengal, and which exposed its wearers to the opprobrious epithet of 'Pandys'. On the representations of the Commanding Officer Lieut.-Colonel S. J. K. Whitehill, it was therefore abolished in 1869, and was replaced by the distinctive badge of a pair of bugles, worn on the collars. Another nickname which long stuck to them was that of the *Bajirao-ki-paltan*, referring to the fact that they had originally been part of the Auxiliary Force of the last Peshwa. A contemporary writer describes them as 'a tall body of men, with a large proportion of Pardeshes'. On January 12th, 1821, the newly constituted battalion received its first colours from the hands of Lady Colville, wife of His Excellency the Commander-in-Chief.[1] The ceremony is thus described by

[1] These colours are still extant. They are in the collection of the late Major S. G. Everitt, to whom they were presented by Mr. Freke Palmer. See Appendix II.

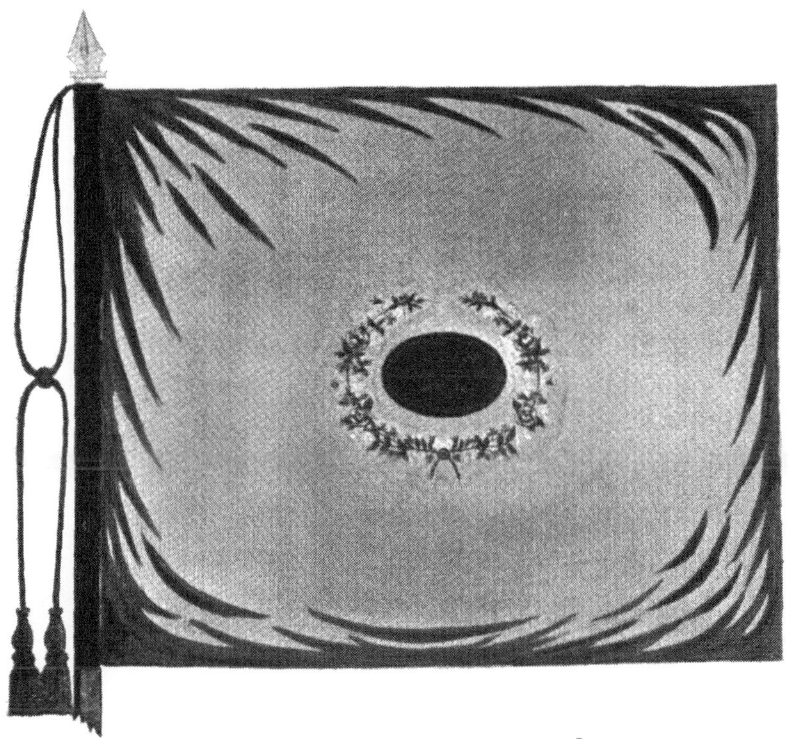

Regimental Colour, Poona Auxiliary Infantry
(1820)

[face p. 22

Lieut.-Colonel Dyson, the Officiating Commandant, in a Regimental Order :—

'It having been deemed requisite by the Commanding Officer of the Poona Brigade that the circumstances attending the presentation of the Colours on the 12th ultimo should stand on the records of the Battalion, the following is published with the Brigadier's hope that the highest flattering compliment of the Colours coming from the hands of Lady Colville will be duly appreciated and have an auspicious influence upon the future fame and conduct of the Battalion, whose conspicuous fidelity at the battle of Kirkee has already been the theme of well-merited praise.

Colonel Elrington's speech : "From the hands of Lady Colville I am proud of the honour of having to present the Colours to the 1st Battalion 12th Regiment of the Honourable East India Company's Bombay Native Infantry, and I have much satisfaction in assuring you, Sir, that I have not only a firm reliance upon their being defended with valour, but that supported by discipline, fidelity, and an heroic devotion to the service, they will in war be ever found forward in the path of glory, and in time of peace become the rallying point of order, harmony, friendship, and hospitality. To you, Sir, and to the gallant band of soldiers under your command, I now consign them, with my best wishes for the honour and prosperity of the Battalion, trusting that mercy to the vanquished may be never lost sight of in any future hour of victory."

Colonel Dyson's reply : "The honour you have this day conferred by presenting the 1st Battalion 12th Regiment Native Infantry with their Colours, would of itself have been a sufficient inducement for this Battalion to imitate the noble example set on all occasions by the Bombay Army ; but this has, in no small degree, been augmented in coming from the hands of the Lady of so distinguished an officer as Sir Charles Colville, our most respected Commander-in-Chief. Allow me, Lady Colville, on behalf of the officers and soldiers of the 1st Battalion 12th Regiment Native Infantry, to return you their sincere thanks, and to assure you they will now anxiously look forward for an opportunity of convincing you how highly they appreciate this honour."

Colonel Dyson cheerfully avails himself of this opportunity of assuring the Battalion, from the orderly conduct evinced, while under his command as well as throughout the march since its leaving Poona, that he feels confident that the expectations of the Brigadier will be fully realized, and that he will be enabled to bring the good conduct of the Battalion to the particular notice of His Excellency the Commander-in-Chief.'

The Attestation Parade, at which the recruit has to swear allegiance with his hand on the Colours, is thus described by a contemporary writer :—

'On the first day of the month, every recruit in the Bombay Army enlisted in the interior was marched to the head of his regiment, and holding in his hand a portion of the colours, took, in his own peculiar dialect, this oath of allegiance :—

"By these colours, I swear I will be faithful and never desert them all my service. I will go wherever I am ordered, and at every place and every time, I will be the faithful servant of the State."' This brief account contains many points of interest : the reference to the 'peculiar dialect' reminds us that enlistment was made regardless of caste, religion, or nationality. Any likely recruit, were he Maratha, Mahar, Surti, Rajput, Sikh, Jew or Christian, was admitted, the majority being usually local men. Apparently there were no difficulties about this arrangement ; the men gave no trouble over caste-questions, the intricacies of which never interfered with duty or discipline, and were well understood and met by the officers, British and Indian. Hence the Bombay sepoy readily swore to go 'wherever ordered', in striking contrast to the pampered Bengali regiments, which objected on caste grounds to crossing, not only the 'black water', but even the borders of the Punjab, Sind or Afghanistan. This accounts for the magnificent services rendered by the Bombay Army in the Sind and Afghan campaigns, in our overseas expeditions to Abyssinia, Egypt and Burma, and also its splendid loyalty to the British Raj in the Mutiny. In 1840, when General John Jacob was looking for picked men for the Sind Artillery, he turned to the Bombay Native Infantry, the sepoys of which he described as 'small and not at all good-looking, but of an amazing energy and activity, and full of zeal and courage, with sinews that no labour could tire, and hearts that no danger could daunt'.

Among the young officers who were now attached to the regiment was Lieutenant James Outram, a 'puny lad' who had come out from England as a cadet in 1819 at the early age of sixteen. His remarkable energy and ability at once attracted notice, and in July 1820 he was posted to the 12th as acting Adjutant. The duties which fell upon this youthful Adjutant were onerous enough, as the following letter, written to his mother from Poona, shows :—

'Many difficulties were thrown in my way, which I had not foreseen. Several officers who were removed from the corps had charge of a company each, all of which were thrown upon my hands, and I had to make out the papers of almost all the companies, besides all the Battalion ones. Almost all adjutants have two writers, one which Government allows—a sergeant—and one which he keeps at his own expense. Now I have been altogether, I daresay, five months without one at all, and have never had more than one at any time. At first a sergeant was not procured (as it is a new corps) till about seven months after I had begun to act. I had, now and then, a writer for a few days, but I daresay I was five months without one altogether ; and when I got the sergeant I found him more a burden than a help to me, as he had everything to learn. . . . I have also been latterly acting quartermaster. I am to be relieved by the regular adjutant,[1] I suppose, on the 1st of next month (May), as he has been relieved from the corps which he has been obliged to remain with till this time. I shall then have done the duties of adjutant exactly ten months.'

Outram threw his heart into his work, and was confirmed in his appointment on January 15th, 1822. His capacity for organization was fully tested when the regiment, which had moved from Poona to Baroda, was ordered to take the

[1] Lieut. R. Ogilby.

field as part of a force sent to quell a disturbance in Mahi Kantha by a rebel chief of the name of Konkaji. At the end of these operations they proceeded, in May 1822, to Rajkot in Kathiawar. At Rajkot the favourite pastime of the officers was pig-sticking, at which Outram was extraordinarily proficient: an old *shikar* book of the period credits him with 74 ' first spears ' out of 123! The time, however, was not entirely taken up in field-sports. Kathiawar was in an unsettled state, as the British Government, though it had inherited the Peshwa's rights in 1818, did not take over the paramount power till 1822. Junagadh was much troubled by troops of disbanded soldiery, which plundered the villages and did much damage until they were suppressed by the left wing of the regiment, under Lieutenant T. D. Morris. While at Junagadh the regiment was inspected, for the first time since its incorporation into the line, by the Commander-in-Chief, Lieut.-General Sir Charles Colville, on February 1824. The Commander-in-Chief in a General Order expressed his gratification at its good composition as a body of men, its smart and soldierlike performance under arms, and its orderly and correct conduct in cantonments—all of which, of course, reflected highly upon the efficiency of the young adjutant, now in his twenty-second year.

In the following month the Regiment moved down to Malegaon in Khandesh, and soon after its arrival, an order by the Governor-General in Council for the reorganization of the Armies in the three Presidencies was received. Twelve two-battalion regiments were allotted to Bombay, twenty-five to Madras, and thirty-four to Bengal. In accordance with this Order, the Army of the Presidency was remodelled and the two battalions of the 12th Regiment became the 23rd and 24th Regiments Native Infantry. Outram was posted to the 24th, but exchanged to his old regiment, where he continued as Adjutant.

Khandesh, at the time of the arrival of the 23rd at its head-quarters, was in a disturbed condition. It had been taken over from the Peshwa in 1818, but had never recovered from the continual inroads of the Moghul and Maratha armies. Bad roads, sparse hamlets, rough impracticable mountain-passes, and jungles in which savage beasts and lawless men roamed at will, were the common features of the country. In March 1825, a formidable rising broke out. The leading rebel, with 800 men, raised the standard of the Peshwa, plundered the town of Anantapur, and escaped with his loot to the hill-fortress of Malair, between Surat and Malegaon. Lieutenant Outram, with 200 men of the 11th and 23rd, and accompanied by Mr. Graham, the District Magistrate, started at once in pursuit. They paraded at 5 p.m. on April 5th and by sunrise had covered thirty-seven miles in seven hours. On the way they could see the fires in Malair, and the surrounding country seemed to be in a blaze. So hot was the pace, that for the last five miles the road was strewn with sepoys, completely exhausted!

Outram had received intelligence, that, if surprised, Malair might be escaladed on the further side. He therefore proposed to carry the place by a *coup-de-main*, to rout the insurgents under the panic of a sudden surprise and, by thus destroying the prestige they had already acquired, to dishearten the allies that were flocking to their standard. This proposition was enthusiastically received by his companions, Ensigns Whitmore and Paul, of the 11th Regiment; but it so far exceeded the discretionary powers which their written instructions vested either in Graham or Outram, that it was a matter of serious deliberation with the former whether he was justified in giving his consent. The result of his inquiries, however, satisfied him that a rapid and alarming extension of the insurrection could only be prevented by offering a prompt check to the rebels. He accordingly sanctioned the proposed measures; and soon after nightfall Outram marched forth to carry them into execution.

'As he neared the hill on which the fortress was situated, he sent Ensigns Whitmore and Paul, with 150 men, to make a false attack in front, while he himself, with the remaining fifty sepoys of his detachment, turning off to the left, proceeded to assail the rear.

'The operation was completely successful. Both parties effected the ascent before daybreak, and while the rebels had their attention drawn to their front by the assault of an enemy whose strength it was impossible to ascertain in the dark, Outram dashed in upon them from behind. The panic-stricken garrison fled with scarcely an attempt at resistance, and at the head of his reunited detachment, and some horsemen whom Mr. Graham had in the meantime collected, Outram followed them up so closely that they could neither rally nor discover the weakness of their assailants. Their leader was cut down; many of his adherents shared his fate, and the rest made for the neighbouring hills in a state of complete disorganization.

'As the infantry had now marched upwards of fifty miles in little more than thirty-six hours, Outram found it necessary to halt them soon after dawn. But the horsemen continued the pursuit as far as the nature of the ground permitted; scouts were despatched to ascertain the point of rendezvous selected by the scattered foe, and, at night, the chase was resumed. The insurgents were a second time surprised; many were slain, numbers were taken prisoners, and the rest, throwing down their arms, fled to their respective villages. A rebellion which had caused much anxiety to the authorities was thus crushed ere the troops intended for its suppression had been put in motion, and the plunder of Anantapur was restored to its lawful owners.'

For this brilliant little exploit, Outram received the thanks of the Commander-in-Chief and the Divisional Commander.

Other work, however, lay ahead. The chief obstacle to the pacification of Khandesh lay in the Bhils. This tribe of aboriginal hillsmen, living in the wild jungles of the Satmala and Satpura hills, had long been a thorn in the flesh of the Marathas, who had hunted them down and treated them with great cruelty. This had driven them into outlawry, their hand against every man, and every man's hand against theirs; they lived by cattle-lifting, marauding and blackmail, and when attacked, retreated to their mountain fastnesses, whence it was almost impossible to dislodge them. The British Government, realizing that nothing could be effected by mere brute force, decided to attempt to reclaim these wild tribes, by encouraging them to settle down as husbandmen, while an appropriate outlet to the energies of the more adventurous

Lieutenant James Outram, 1819
Adjutant, 23rd Bombay N.I.
(Courtesy of John Murray)

[*face p.* 28

spirits was to be provided by the formation of a Bhil Corps. For this purpose, the Bhil district was divided into three Agencies, and in 1825 Lieutenant Outram was seconded from his regiment to take charge of the North-Eastern Agency, with the special duty of raising the Corps.

But it was first necessary to convince the Bhils that their strongholds were not impregnable, and on one occasion Outram, taking over a detachment of thirty men of the 23rd from the Indian officers in charge of them, made a daring attack at daybreak upon a party which, under one Pandu and other local leaders, had long been a terror to the district. The Bhils fled panic-stricken, leaving their women and children in the enemies' hands. The 23rd, under Major Deschamps, then engaged in a regular drive against their haunts, which had such a salutary effect that Outram was able to set about his task of enlisting men for the Corps. At first it was intensely difficult to induce these shy, suspicious jungle-folk even to come near a British encampment, but gradually Outram, by venturing fearlessly and alone into their most impenetrable jungles, by his skill as a *shikari*, and above all by his prowess as a tiger-slayer, won their confidence. In this respect, his record was an extraordinary one. During his ten years among the Bhils, he slew 235 tigers, besides bears, leopards and buffaloes ; one of them he followed up for three miles on foot and finished off with a hog-spear, accompanied by six Bhil trackers armed only with bow and arrow. Another which had got him down and was clawing him, he shot dead with his pistol, and when his faithful followers raised a cry of alarm, he silenced them with the remark, ' What do I care for the clawing of a cat ? ' On one occasion when a tiger was hiding in a ravine, he made his followers lower him from an overhanging branch in a kind of sling made from their turbans, and shot the beast dead while dangling in the air ! By these means he obtained an extraordinary ascendancy over them, and eventually persuaded five of the most adventurous to join him as a body guard. One by one the number increased to sixty. At last, in January 1826, Outram triumphantly marched into Malegaon at the head of 134 Bhils, and success was assured when the sepoys of the 23rd rose nobly to the occasion, and showed themselves worthy of their leader. Laying aside the natural repugnance of the caste Hindu for the jungle tribes, they received them into the cantonment with open arms, offering them betel-nut and other tokens of friendship. In Outram's words, ' not only were the Bhils received without insulting scoffs but they were even welcomed as friends, and with the greatest kindness, invited to sit down among them, fed by them, and talked to by high and low—as on an equality from being brother soldiers. This accidental circumstance will produce more beneficial effects than the most studied measures of conciliation ; and Bhil reformation will owe much to it. The Bhils returned quite flattered with their reception, and entreated me to allow them no rest from drill until they became equal to their brother soldiers.'

The prophecy was quickly fulfilled. Recruits flocked in, and by the following year, the Bhil Corps was actually taking the field against its former brethren, and doing invaluable work, as no one knew the jungle as they did. By 1828 the Collector was able to report that, for the first time for twenty years, the land had enjoyed six months' rest. Like Nicholson and other great Englishmen in India, Outram was posthumously deified. Many years after his death some of his old sepoys, coming by chance across an ugly little image in which they fancied they saw some resemblance to the rugged features of their beloved leader, forthwith set it up and worshipped it as ' Outram Sahib '. His actual connexion with the Regiment ceased from this time, though he continued to be borne upon its books until 1853, and it was named after him in 1903. Writing about Outram as he knew him in those early days, his old comrade-in-arms, Major Giberne, said : ' He was about six years my senior, and I well remember the kind of romantic deference with which the younger officers all regarded him. Even as a young man, he was possessed of that rare combination of sound judgement, quick decision, unflinching determination, calmness both in counsel and in action, and fearless intrepidity, which so pre-eminently distinguished him in after years, and which, while he himself invariably led the way on every occasion of doubt, difficulty or danger, seemed instinctively to command the sympathy and support of everyone, whether European or native, who followed his leadership or acted under his orders.' His influence upon all with whom he came into contact was magical, and it has been rightly said that 'the spirit breathed into the regiment by Outram has lived on up to now'.

AUTHORITIES

James Outram: a Biography. By Major-General Sir F. J. Outram. 2 vols. London (1880).

The Bayard of India. By Captain L. J. Trotter (1903).

Services of Lieut.-Colonel Outram (1853).

Outram Letters.

Margaret Outram, 1778-1863. By M. F. Outram (1933).

Tribes of Khandesh. By Major D. C. Graham (1845).

A Memoir of the Khandesh Bhil Corps, 1825-91. By A. H. A. Simcox (Bombay, 1912).

Dictionary of National Biography, article on James Outram.

Gazetteer of the Bombay Presidency, Volume XII (Khandesh). Regimental Records.

Capt. W. A. M. Wilson's *Historical Records of the 23rd Regiment*.

EARLY YEARS: THE SIND CAMPAIGN

CHAPTER III

(1825-43)

EARLY YEARS: THE SIND CAMPAIGN

AFTER spending two years on detachment duty in Khandesh, the Regiment was transferred to the famous fortress of Asirgarh, which may be seen to-day from the railway, towering above the surrounding country to the height of 2,300 feet. It had been captured by General Doveton in 1819, and, standing as it does on the main line of communication with Central India, it had a garrison down to 1901. Life at Asirgarh was uneventful, the only occurrence of note being the death of Lieutenant Barlow, on August 3rd, 1827. Lieutenant Barlow was an accomplished linguist, who had joined the Regiment in 1820, and had succeeded Outram in 1825 as Adjutant. A striking proof of the *cameraderie* existing between officers and men is shown by the fact that the sepoys, laying aside all thoughts of caste, volunteered to carry his body to the grave. In 1829 the regiment moved to Baroda, and in 1833 to Satara, where they found themselves once more under the command of an old comrade-in-arms in Colonel Lodwick, who had been Brigade Major at Kirkee. That the Regiment had used these long years of peace to good purpose, is apparent from the General Order issued by the Commander-in-Chief, Sir John Keane, after inspecting them on November 25th, 1835, in which he states that 'the splendid appearance of the 23rd Regiment, Native Infantry, which possesses a fine body of men, attracted the Commander-in-Chief's peculiar notice, and their movements in the field were equally satisfactory, a proof to His Excellency that Major Wilson[1] has bestowed much attention upon his duties, and was

[1] Major, afterwards Lieut.-Colonel, G. H. Wilson joined the Indian Army in 1805, was posted to the 23rd on June 7th, 1824, and commanded it from February 5th, 1833 to January 4th, 1840, when he was transferred to the 14th N.I.

rewarded by finding himself at the head of the Regiment of which he has reason to be proud.' Writing privately to Major Wilson, the Commander-in-Chief paid him the high compliment of saying that 'it would be difficult to find a corps in more creditable order', and directing that 'his best and warmest thanks' should be conveyed to every officer and man.

In November 1836, orders were received for the Regiment to proceed to Bombay, and on November 5th, the anniversary of the battle of Kirkee, Colonel Lodwick, commanding the Satara District, issued the following Order:—

'On the movement of the 23rd Regiment Native Infantry, after being stationed at Satara nearly four years, Colonel Lodwick performs a pleasing duty in recording his sense of its conduct during that period. Associated with this Regiment in the glorious action of which this day is the anniversary, and witness of its well-earned claim to be admitted on the Field to the honour of entering the ranks of the Regular Army, Colonel Lodwick has ever felt a lively interest in its career, and after a lapse of nineteen years it affords him the highest gratification to be able to express his opinion that the conduct of the 23rd Regiment Native Infantry, while stationed at Satara under his command, has been exemplary in all the qualities becoming good soldiers, during a period of profound peace, as it was marked by steadiness and gallantry in the day of battle.'

On November 10th, 1836 the Regiment marched for Bombay, arriving there on the 21st. During this year an insurrection broke out in Coorg, and the rebel tribe of the Gandas entered the town of Mangalore, burned the public offices, and then retreated. On April 12th the Regiment was ordered on Field Service to aid in quelling the insurrection. One hundred and fifty rank and file, under Captain Newport, embarked on the evening of April 14th, on board the H.E.I.C.'s sloop-of-war *Amherst*; the Head-quarters and the remainder of the regiment embarked the following morning on the H.E.I.C.'s ss. *Atalanta*, and arrived off Mangalore on the evening of April 18th. The detachment on board the *Amherst* was disembarked at Mangalore, and the *Atalanta* proceeded to Causergode further south, where the Regiment disembarked on the night of the 21st, and encamped near the village. From there a detachment of 150 men under Captain Scott was sent to occupy the village of Vithul on May 5th, and on the 10th the Head-quarters marched for Mangalore, leaving behind at Causergode a detachment of a hundred men under Brevet Captain Liddel. The Head-quarters reached Mangalore on May 12th, and on the 17th and 18th embarked on the transport *Columbia*, and arrived in Bombay on June 2nd. Captain Liddel's detachment at Causergode was embarked on the *Amherst*, which was driven by stress of weather to Trincomalee and eventually went to Madras! The men were disembarked and Lieut.-General Sir P. Maitland highly complimented Captain Liddel and 'this detachment for their soldier-like appearance after the hardships they had endured from a crowded ship, stress of weather and want of food'.

In 1838 the First Afghan War broke out. Lord Auckland, the Governor-General, decided to invade Afghanistan, depose the ruler, Dost Mahommed, who was suspected of pro-Russian tendencies, and replace him by Shaha Shuja, who had been driven into exile about twenty years previously, and had been living in the Punjab as a British pensioner. The invasion of Afghanistan was to be carried out by two forces, the Bengal Army, 14,000 strong, advancing from Firozpur, and the Bombay contingent, about 5,000 strong under Sir John Keane, which was to land at the mouth of the Indus, and proceed up the river as far as Sukkur. It was then to strike across the desert to the Bolan Pass, and effect a junction with the Bengal army at Kandahar. The operation was bristling with difficulties, and violated all the conditions of sound strategy. The Baluchi tribes were hostile, and the Mirs of Sind were bitterly opposed to the occupation of their country, which they regarded as a direct violation of the treaty of 1832. Moreover it involved a hazardous march across the formidable Sind Desert. The Regiment received orders to proceed on active service in Sind on September 15th, 1838. It embarked on the war steamer *Semiramis* and the *Palinurus* sloop on the 22nd, under Major Wilson, and after what seemed to be interminable delays, finally landed at the Hazamro Creek, about seventy miles south of Karachi on December 6th, where it became part of the second Infantry Brigade, consisting of H.M. 17th Regiment, and the 19th and 23rd Native Infantry. On December 28th the Brigade marched to Tatta, where it received the complimentary remark from the Commander-in-Chief that its 'appearance and regularity on the line of march was creditable to veterans'. At Tatta, further delays were experienced, as the Mirs were putting all kinds of obstacles in the way, and Outram, who had volunteered to rejoin his old regiment for active service, was sent instead to Hyderabad to treat with them. Finally, however, the army moved by slow stages to Larkhana in Upper Sind, where the troops were reorganized. The 23rd, the 1st Bombay Grenadiers and the 5th Bombay Native Infantry now became the 2nd Infantry Brigade of the Bombay Division of the Army of the Indus. On March 11th and 12th, the main body left Larkhana for Afghanistan, and on the 1st the 2nd Brigade started for Sukkur, which was reached on the 21st.

On May 29th a wing of the 23rd, consisting of 300 rank and file under Major Newport,[1] with the 42nd Native Infantry and some irregular cavalry, was called upon to form an escort to a convoy of 4,000 camels laden with grain, stores of all kinds, and ten lakhs of rupees for the Army of the Indus, which had to be despatched across the desert from Shikarpur to Dadar, at the eastern end of the Bolan pass, a distance of about 170 miles. It was now the height of the hot weather; villages were few and far between, water scanty and brackish, and forage almost unprocurable. The convoy was impeded by endless multitudes of followers and convalescents going up to join their regiments. Major-General Seaton in his book, *From Cadet to Colonel*, vividly describes the horrors of this memorable march. At noon the sun was directly overhead, throwing no shadow, and without a cloud in the sky to moderate the fierceness of his heat. The thermometer stood at 117° in the shade.

'In the evening of the 29th we struck camp, and commenced our march at sunset. As soon as we entered into the

[1] Major C. Newport succeeded Lieut.-Colonel Wilson as Commandant in 1840 and retired two years later.

UPPER SIND AND BALUCHISTAN

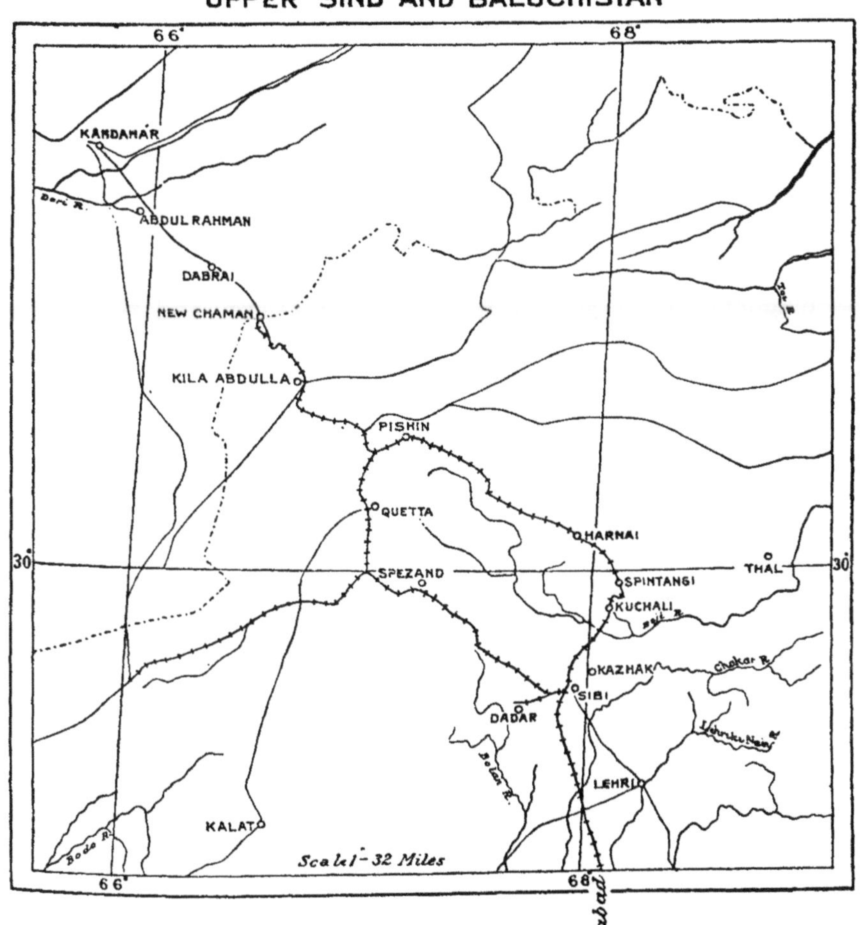

Desert a wind sprang up, gentle at first, then hot and fierce, bringing with it particles of dust, fine as the finest powder, which penetrated everything, and, with the heat still radiating from the soil, created an intolerable thirst. The sepoys, each with his heavy musket, sixty rounds of ammunition, clothing, haversack with necessaries, accoutrements and his brass pot filled with water, were heavily laden for such a march, the burden doubling the already unbearable oppression of their tight-fitting woollen uniforms. The condition of the men in such circumstances was pitiable, and every minute their sufferings increased.

'The water in the men's brass pots was soon exhausted, for the hot wind and the dust, as I said before, created an intolerable thirst, and they drank without restraint. At midnight they began to flag, then to murmur, and shortly there was a universal cry of "Water! water!" Major Newport halted, put a strong guard over the water as fast as the camels came up, and it was then served out. Such was the eagerness of the poor creatures that a frightful tumult arose. The camp-followers, in the hope that they might succeed in obtaining by some lucky chance a draught of water, rushed in among the sepoys. But the guard kept the people back. A sergeant was then appointed to serve it out, and so great was the distress which many suffered from the want of it, that even the all-powerful prejudices of caste were forgotten, and Hindus drank out of the leathern bags, in common with the Mahomedans, water served out by a European sergeant.

'Each sepoy came up in succession and received his portion of water. Some were in despair at the smallness of the quantity, many were even in half-raving state, while a few, whose sufferings had not been so great as those of their comrades quietly took the allowance that was handed to them.

'But the poor, heavily-laden camp-followers, some carrying infants, were in a pitiable condition, and the children's cries were heartrending. Strong men, exhausted by carrying loads, were scattered on the ground, moaning and beating their breasts; others lying down quite exhausted.

'After an hour's halt we started again and pushed on faster and, most fortunately, the burning wind gradually subsided.

As everyone felt the necessity of water, the camels were urged on, but they could go little beyond two and a half miles in the hour. Towards daybreak our line of march lay parallel to a ravine. I rode a little distance from the column, and followed the ravine in all its windings, in hopes of finding water, as I observed several moist places. A sepoy, who skirted it with me, was in such a state that, when I spoke to him, he could scarcely reply; his tongue rattled in his mouth, and his whole countenance was distorted with agony. I had read of such suffering, but had never witnessed it until now.

'Day dawned upon the frightful waste—a boundless plain of hard alluvial soil, apparently deposited by the annual overflow of the Indus. The ravines we met with were the channels cut by the retiring waters. Not a tree, bush, shrub, or blade of grass was to be seen—nothing but a scene of dreary desolation; and the road over this horrible plain was distinctly marked by the skeletons of men, camels, and horses abandoned by kafilahs, or by the army that had preceded us. Wherever a foot trod, the surface of the deposit was broken into an impalpable powder, in which there was something peculiarly irritating, and which the lightest wind carried aloft. It was this fine dust settling on the clothes and sticking to the skin that, united with the hot wind, created the thirst that so overpowered everyone.

'At sunrise, I rode ahead of the advance guard, and in the horizon I espied the towers of the little fort where we were to halt. I galloped back to proclaim the joyful news, which spread like wildfire. The men and camp-followers cheered up, and even the cattle seemed by a sort of sympathy to understand that relief was at hand, and brightened up a little. At our exhortation, all pressed on, and the people ceased talking, everyone doing his best to reach the halting-place before the sun got high.

'At length the terrible march was over. Immediately the men saw the wells, discipline was at an end, and before the guards could be told off, the eager multitude made a rush for them, and commenced drawing water, which, notwithstanding its disagreeable taste, they drank in thoughtless haste.

4

'At this place, out of thirty-two wells dug in the bottom of a ravine, only six contained water. One of them was poisoned by an animal which had fallen into it, and of the others, the water was so bitter and brackish that the men said it turned their lotahs black.

'The name of this fort was Burshoree, and here we halted for a day. Round the place was a little cultivation; and, amongst other things, some fields of melons and cucumbers. The people flew at these and ate immoderately to quench their raging thirst, for guards had now been put over the wells, the quantity of water in which was so limited, that it had to be served out to officers and men alike, a quart at a time. The consequences of this inconsiderate conduct of the men soon appeared in the fearful form of cholera.

'We marched at ten o'clock on the night of the 31st. It was my tour of duty to command the rear-guard, and, on going over the ground with a small party to hurry off the loiterers, I found several camels dead, with their loads lying beside them, and three poor wretched surwans (camel drivers) in the last stage of cholera, dying, deserted by their comrades. I could do nothing for them. I had no medicine, no brandy—nothing, in fact; so I sent for the head man of the village, and made the poor creatures over to his charge, with instructions how to treat them. It was all I could do.

'At the place where we should have encamped the water was dried up. In order, therefore, to save the men from unnecessary fatigue, the column was halted, and a party of horsemen sent on with two of the guides to search for water in several places where it was anticipated it might be found. After a very long delay, the horsemen returned with a report that water had been found at a place called Hadjee ka Chouk, four miles farther on. This unavoidable addition to our march was attended with fatal consequences. From where we halted to Hadjee ka Chouk, the route of the column was marked by scores of men scattered along the road, ill and dying from fever, cholera, and sheer exhaustion. A march in our own provinces, during the hot winds, is bad enough, but, compared with this, it is pleasant. The sun was high, and the wind, fiercely hot, seemed not only like fire, but stuffy—stifling, as if it could not satisfy the lungs, a feeling caused, I suppose, by the fine impalpable dust with which it was laden. The distress caused by the heat and want of water was terrible, and the cries and entreaties of the poor wretches who, as they sank by the road-side, saw the rear-guard passing, and feared they would be left to die, were most heartrending. Many never rose from the ground where we halted, but quietly died there; others, stricken with fever, struggled on a little farther.

'I cannot describe our sufferings from the heat, the dust, the desert wind, the myriads of flies, and the stench of the dying and dead camels which rendered our lives intolerable. The heat in our tents rose to 119 degrees; the whole camp smelled like a charnel-house, and in very truth it might be called one, for no person could take three steps in camp, any-where, without seeing a dead or dying man or animal. There were a few villages near our camp, this day, with a sprinkling of trees and stunted bushes, and the country was not so desert like, but it had no effect in mitigating the sickness.

'During the day there was a cry of "Belooch, Belooch—they are carrying off our camels". Dressed as we were, in shirt and loose cotton drawers, and with our thick hats on, we snatched up our arms and rushed out, all who were able hurrying off with the nearest picket, and arrived just in time to support some of our troopers who were keeping the marauders at bay. We saved our camels and the Beloochees went off, some of them severely wounded. The loss of our camels would have been destruction to the convoy.

'The sun, still high above the horizon, looked like a ball of red-hot copper. The fine dust of the desert, driven across its face by the howling wind, sometimes in streaks, sometimes in clouds, gave to everything a sickly, unearthly appearance; and with little effort of the imagination disease and death might be supposed to be flying on its wings, cutting us off from the pure breath of heaven beyond, and from all hope of escape from this frightful charnel-house.

'The heat was as terrible as ever, and the strongest amongst us began to droop. Not only were our actual bodily sufferings great from this cause, and from want of rest and sleep, but our minds were tortured by what we saw around us, and by the fears we naturally entertained for the possible consequences that might ensue to the whole convoy. The thermometer still ranged to 119 degrees in our tents—a heat so fierce that it utterly prevented sleep during the day, and at night, if we did not march, we were kept awake by the groans of our dying, by the roaring of the camels, or by the necessity of attending our sick comrades, and satisfying the increased demands made by duty; for as death thinned our ranks or sickness disabled us, an increase of duty fell on our diminished numbers. On the night of the 5th we marched again, and in the morning reached the little walled town of Baugh, and the passage of the Desert was accomplished.

'We found a detachment of our troops occupying the town, and got some medical aid. All help in the shape of carriage, cattle, and water was sent back to the last dreadful camp, to bring up the sick and dying. The best doctor of all was the complete change of scene, which cheered up everybody. The sight of gardens, green fields, and cultivation, shut out the hideous vision of the Desert, and tranquillized men's minds. The cooler air, though the difference was but a few degrees, and abundance of clear pure water, refreshed us all; and as with these blessings we could now enjoy that rest and sleep which we so much required, sickness soon disappeared from our camp.

'We marched on the morning of the 7th, four miles north-east of Baugh, to an immense pool of water surrounded by low jungle, with villages and fields scattered about the country. The sepoys and camp-followers rushed into the water with shouts of joy, and from that moment everyone began to mend. In those seven fearful days we had lost six out of the fourteen Europeans in camp, and a hundred sepoys. More than three hundred camp-followers were known to have perished.

'We halted two days at the pool, and as the people had in a great measure recovered, we marched on the 10th at night. We had no sooner moved off the ground than a large party of Beloochees made a dash at our camels, but were very speedily and sharply repulsed, for our men were now quite up to their work, all the sick having been left behind at Baugh. As we drew nearer to the hills, these attacks became more frequent, and though we lost a few camels, the Beloochees in several engagements were handled very severely, and our cavalry always brought in a few of the marauders' heads.

'On the 12th we reached Dadar, a small walled town at the entrance to the celebrated Bolan Pass. As we approached the town we crossed two beautiful clear sparkling streams running between fields of cotton plants in full bloom. The sight of the water and the odour of fresh fields revived our spirits so much that we felt as if we had stepped out of the valley of Death into the garden of Eden, and we continued our march with renewed hope and in good spirits.'

On the arrival of the Detachment at Dadar on June 12th, the following order was issued by Brigadier Gordon, Commanding Upper Sind. 'The Brigadier has received the official report from Major Newport announcing the arrival of the valuable convoy, under his charge, at Dadar. The Brigadier avails himself of this opportunity to record thus publicly his thanks to the Major for his indefatigable and zealous exertions in conducting a khafila of nearly four thousand camels through a country which, even at the most favourable period of the year, presents difficulties to a military body of no ordinary nature, and at this season exposed to a burning sun, the powerful effect of which it was scarcely possible for human beings to contend against. Added to which the great want of water was sadly felt, besides being harassed at all points by marauders from the enemy.

'These convoys with large amounts of treasure were of the most urgent necessity for the Army in advance, and the cheerfulness and alacrity with which the whole duty has been performed so very creditable to the officers and men, and it will be the subject of report to the high authorities under whom Brigadier Gordon is acting.'

In August 1840, the Regiment was at Sukkur, a detachment of two hundred rank and file having been despatched to Dadar. On October 29th, 1840 Dadar was attacked in force by Nasir Khan, the heir of the Khan of Kelat, with about five thousand Brahuis. The enemy, aided probably by sympathizers inside, succeeded in destroying part of the town, but they were unable to penetrate into the entrenched camp which had been formed by the detachments of the 5th Bombay Native Infantry and the 23rd under Captain Watkins.[1] The little force behaved with great gallantry and steadiness. The attacks went on almost continuously for three days, but despite immense superiority of numbers on the part of the enemy, they were beaten off with heavy loss. Dadar was a post of great importance on the line of communications as a depot for stores to be sent through the Pass, and in a despatch addressed to the Political Agent in Upper Sind, the Secretary to the Government of India observed that 'the steady firmness and untiring bravery manifested by the small body of troops under Captain Watkins in the defence of their position at Dadar against the force opposed to them for several days has elicited the high admiration of Government'. The detachment, in February of the following year, marched to Quetta as escort to Mr. Ross Bell, the Political Agent, who went there in order to settle the Kelat succession, and on April 26th the whole regiment returned to Sukkur. On May 10th, 1841, in recognition of their good work in Sind, an order was received that 'the 5th and 23rd Regiments, Native Infantry, be organized as Light Infantry, and styled respective the 5th and 23rd Regiments Native Infantry (or Light Infantry)'. On October 31st, 1842, the Regiment set out for Karachi, and after halting for some time at Bhuj, eventually embarked at Mandavi on H.E.I.C.'s frigate *Auckland* for Bombay, which was reached on February 3rd, 1843, after an absence from the Presidency on Field Service for three years and four and a half months.

AUTHORITIES

Regimental records. Captain W. A. M. Wilson's *Historical Records of the 23rd Regiment*.

The account of the famous desert march is taken from Major-General Sir Thomas Seaton's *From Cadet to Colonel* (1877).

[1] Captain, afterwards Major, J. Watkins joined the Regiment in 1824, and commanded it from November 1st, 1847 to his death on June 10th, 1854.

SAVANTWADI, THE PERSIAN WAR AND THE INDIAN MUTINY

CHAPTER IV

(1843-1859)

SAVANTWADI, THE PERSIAN WAR AND THE INDIAN MUTINY

THE 23rd did not long enjoy their well-earned repose. The Southern Maratha Country had long been in an unsettled condition, and serious risings broke out in several districts in 1844. These mostly originated in the little State of Savantwadi, where the turbulent nobles who had never acquiesced in the transfer of their country to British rule were perpetually creating trouble, aided by malcontents from Portuguese territory. Lieut.-Colonel Outram, now Political Agent at Kolhapur, proceeded to the scene of the outbreak with the 10th Bombay Native Infantry, but the wild, wooded character of the terrain made operations very difficult. Marauding parties under rebel chiefs invaded British territory, burning villages and collecting revenues. By 1845, the whole countryside was in disorder. Outposts were constantly attacked, and even close to military forts, there was no security of person or property. Attempts were made in vain to win over the Indian officers. A force of 1,200 men was now despatched to the scene of the rebellion and placed under Outram's command; this contained the Left Flank Company of the 23rd, one hundred strong, under Lieutenant Peyton. The focus of the rebellion was the stronghold consisting of the twin forts of Manohar and Mansantosh, which tower up to the height of 2,500 feet above the plain at the extremity of the Ragna Pass, about a mile from the Ghats, and thirty-five miles from Vengurla. They are separated by a yawning chasm, two hundred yards wide. Outram advanced upon his objective at lightning speed, the troops hewing their way through pathless jungles, ascending steep passes, and demolishing stockades set up by the rebels, until, on January 24th, they found themselves at the foot of Mansantosh. Outram now determined to divert the enemy's attention by a feint attack on Manohar, while the detachment of the 23rd, under Peyton, rushed the stockades protecting Mansantosh. 'Having determined on storming the stockades,' he says in his despatch, 'lest longer delay should strengthen them still further, I sent Lieutenant Peyton with the company of the 23rd Regiment to occupy a belt of jungle running up the scarp of the fort to the left of the stockades, with orders to ascend till within forty or fifty yards of the scarp, where the cover was sufficiently dense to shelter his men from the stones hurled from the fort above, or shot from the stockades in flank.' Unfortunately, owing to the difficult nature of the ground, the troops had not reached their positions when the advance sounded. 'Relying, however,' Outram continues, 'on the gallantry of the troops composing the storming party, and feeling the ill-effect of the further delaying to take the stockades, I sounded to the stormers to throw out skirmishers to the left. Lieutenant Gardiner gallantly led, thus turning the flank of the enemy's position. The whole steadily ascended the steep ridge, at the top of which a succession of three stone stockades were occupied by about 150 of the enemy, who opened a heavy fire upon the stormers, also exposed to showers of stones from the top of the fort immediately over them. The stockades were carried with little further difficulty than that of climbing the very steep ascent, and the enemy fled the moment they saw their flank turned.' During these operations Lieutenant Peyton was severely wounded, and his place was taken by Captain Watkins. After an unsuccessful attempt to blow in the gate of the fort, it was determined to send a small storming party to rush it under cover of darkness; but this was forestalled by the enemy, who evacuated both forts during the night, and managing to evade the troops stationed at the foot to intercept them, made off for Goanese territory. Outram at once started in pursuit, and the enemy were severely handled before they managed to cross the border, but it was not until April 15th that the spirit of the revolt was broken, and the rebel chiefs began to sue for pardon. On May 12th, the Left Flank Company rejoined Regimental Head-quarters. Meanwhile, however, the trouble had spread to the Northern Konkan. Accordingly on January 7th the 23rd marched to Thana, where Major Scott assumed command of a Field Force consisting of the 23rd, the 15th, the 48th Madras Native Infantry, and the Native Veteran Battalion. This column was constantly on the move in the disturbed areas of the Konkan until the rains, when it took up its quarters at Nasik, leaving a strong detachment at Thana. It did not return to Bombay until February 1846, when it received the thanks of the Court of Directors for its services in these trying operations.

From 1846 to 1856, the 23rd enjoyed what was a very rare occurrence in its early annals—a period of ten years uninterrupted peace, which was at length disturbed by the outbreak of war with Persia. For over a year relations had been unsatisfactory: the Persians, in violation of all existing treaties, continually attempted to occupy Herat, which had been a bone of contention ever since 1838. The British Government was extremely nervous about this place, the strategic importance of which was greatly exaggerated, and were apprehensive lest it might eventually fall into Russian hands. The arrogance and duplicity of the Persian Court and their marked discourtesy towards the British mission at Teheran made hostilities inevitable, and in November 1856 a combined force was sent under Major-General Stalker to occupy the island of Kharak and the city of Bushire in the Persian Gulf. Bushire was bombarded and captured, and in January 1857 a second division was despatched to the scene of operations to reinforce the Expeditionary Force under Outram, now Major-General Sir James Outram, K.C.B. The 23rd was to form part of the 2nd Infantry Brigade of this division, the other battalions being the 26th, and a special Light Battalion composed of ten Light Companies from various Bombay regiments. Orders for embarkation found the 23rd at Rajkot, under the command of Major Travers. Movements of troops were very slow in those days. Although the Regiment left Rajkot on January 5th, it was not until February 9th that it embarked at Porbander on the transport *Saldanha*, its strength being twenty British officers and five hundred Indian troops of all ranks. Unfortunately, the transports were delayed by tempestuous weather in the Persian Gulf, and did not arrive in time to take part in the brilliant little action of Kush-ab, where the Persian army, 11,000 strong, was completely defeated, the Bombay Cavalry achieving the rare distinction of riding down a battalion of unbroken infantry and cutting it to pieces. Outram then returned to his entrenched camp outside Bushire, to await the arrival of the 23rd, after which he proposed to leave a garrison at Bushire under General Stalker, while he himself with the remainder of the Army, about 4,800 strong, landed at the delta of the Euphrates. It was feared that the long, rough sea-voyage of over a month would have rendered the sepoys temporarily unfit for further exertions, but Outram was agreeably surprised to find that this was not the case. Writing to Lord Canning, he says: 'Supposing they might be almost incapacitated physically, after so long an absence from cooked food, from satisfactory working in active service immediately on landing at Mohumra [Muhamra], I had arranged to substitute an equal strength of Native Troops from General Stalker's Division for the 23rd Regiment. But the men are so eager to go on, and are really so strong, declaring that two days' cooking on shore here will fit them for anything, that I have decided on taking them.'

After a few days ashore to refit, the 23rd re-embarked on the *Saldanha* on March 15th, arriving at the Shat-ul-Arab on the 23rd. On the following day the attack on Mohumra began, and the ships moved up the river to silence the batteries at the junction of the Karun and the Euphrates. These operations, so strangely anticipating those undertaken by another Bombay army in November 1914, are best described in Outram's despatch to H.E. the Commander-in-Chief, dated March 27th, 1857:—

'Sir,—I have the honour to report for your Excellency's information the successful result of the operations against Mohumra. The Persian Army evacuated their entrenched position and camp yesterday about mid-day, leaving behind all their tents standing, with nearly the whole of their property, public and private, all their ammunition, and seventeen

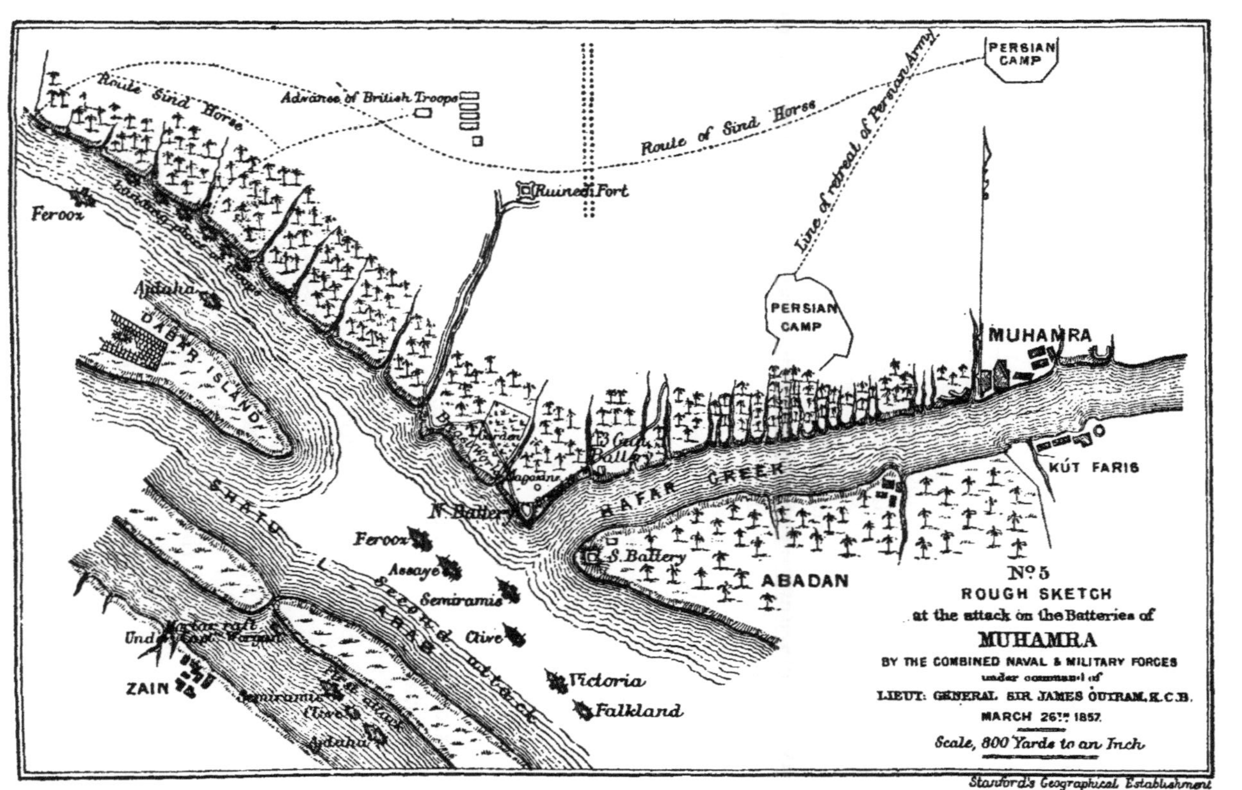

Nº 5
ROUGH SKETCH
at the attack on the Batteries of
MUHAMRA
BY THE COMBINED NAVAL & MILITARY FORCES
under command of
LIEUT: GENERAL SIR JAMES OUTRAM, K.C.B.
MARCH 26TH 1857.
Scale, 800 Yards to an Inch
Stanford's Geographical Establishment
PERSIAN CAMP
PERSIAN CAMP
MUHAMRA
KÚT FARIS
ABADAN
HAFAR CREEK
S. Battery
N. Battery
Ruined Fort
Route Sind Horse
Route of Sind Horse
Advance of British Troops
Line of retreat of Persian Army
DABAR ISLAND
SHATU - L - ARAB
Second attack
First attack
ZAIN
Ajdaha
Feroox
Feroox
Assaye
Semiramis
Clive
Victoria
Falkland
Semiramis
Clive
Ajdaha

guns—as far as I am able to learn, only five guns and a portion of the personal effects of the Shah Zada have been carried away.

'It was my intention to have proceeded against this place immediately upon my return from the Borazjoon expedition last month, but owing to the non-arrival of the requisite reinforcements, caused by the tempestuous weather in the Gulf, together with unforeseen causes of delay, I was not able to have Bushire until the 18th instant.

'For some months past the Persians had been strengthening their position at Mohumra; batteries had been erected of great strength—of solid earth, 20 feet thick and 18 feet high, with casemated embrasures—on the Northern and Southern points of the banks of the Karun, and Shat-ul-Arab, where the two rivers join. These, with other earthworks, armed with heavy ordnance, commanded the entire passage of the latter river, and were so skilfully and judiciously placed, and so scientifically formed, as to sweep the whole stream to the extent of the range of the guns up and down the river, and across to the opposite shore; indeed, everything that science could suggest, and labour accomplish in the time, appeared to have been done by the enemy, to effectually prevent any vessel passing up the river above their position. The banks for many miles were covered by dense date groves, affording the most perfect cover for riflemen; the opposite shore being neutral territory (Turkish) was not available for the erection of counter-batteries.

'After mature deliberation, I resolved to attack the enemy's batteries with the armed steamers and sloops of war; and, so soon as the fire was nearly silenced, to pass up rapidly with the Troops in small steamers, towing boats, land the force two miles above the northern point, and immediately advance up and attack the entrenched Camp.

'I have now the very great satisfaction of announcing to your Excellency the complete success of the first two operations; the third, to the regret of the Army, being frustrated by the precipitate flight of the enemy.

'The Persian Army was ascertained from credible reports to amount to 13,000 men of all arms, with thirty guns, was commanded by the Shah Zada Prince Khamler Meerza in person.

'The British force under my command composed as follows,[1] was the utmost I deemed it prudent to withdraw from Bushire; but with the aid of four armed steamers and two sloops of war to effect my landing, I felt confident of success, although I anticipated some loss from what I learnt of the determination expressed by the enemy to oppose our further advance to the utmost of their power, and their extreme confidence of succeeding as evinced by the fact of their having sent away their baggage cattle.

'On the 24th instant the steamers with transport ships in tow, moved up the river to within three miles of the Southern Battery, but as some of the large ships shoaled on the way, and did not reach the rendezvous till after dark, I was obliged to defer the attack for another day. During the night a reconnaissance was made in a boat to ascertain the nature of the soil on an island west of, and immediately opposite the Northern Battery, where I wished to erect a Mortar Battery, but as it was found to be deep mud, I determined to place the mortars upon a raft.

'The horses and guns of the Artillery, a portion of the Cavalry and the Infantry, were transhipped into boats and small steamers during the day, in readiness for landing the following morning.

'At break of day on the 26th the mortars opened their fire upon both the Northern and Southern Batteries. At seven o'clock the several vessels of war moved up into the positions allotted them by Commodore Young, and by 9 o'clock the fire of the heavy batteries was so reduced that the small steamers with boats in tow, were able to pass up and land the Troops above the Northern Battery.

'By half past one o'clock the Troops were landed, and formed and advanced without delay through the date groves and across the plain upon the Entrenched Camp of the Enemy,

[1] H.M. Light Dragoons 89, Sind Horse 303, H.M. 64th Foot 704, 78th Highlanders 830, 23rd Regiment, Bombay Native Infantry 749, 26th ditto 716, Light Battalion 920, Bombay Sappers and Miners 109, Madras ditto 124, 12 guns, 3rd troop Horse Artillery 166, No. 2 Light Field Artillery 176. Total 4,886.

who, without waiting for our approach, fled precipitately, after exploding their largest magazine, leaving their tents and baggage, public and private stores with several magazines of ammunition and sixteen guns behind. The want of cavalry prevented me from pursuing them as I could have wished, but I despatched a part of Sind Irregular Horse, under Captain Malcolm Green, to follow them up for some distance. This officer reported that he came upon their rearguard retiring in good order, but that the road in many places was strewed with property and equipments. The loss of the Persians has been estimated at two hundred killed, among whom was an officer of rank and estimation, Brigadier Agha Jan Khan, who fell in the northern battery.

'With the exception of the artillery, with the mortar-battery under Captain Worgan, no portion of the military force was actively engaged with the enemy, beyond some European riflemen sent on board the war-vessels, but I am not the less indebted to all for their exertions and zeal, especially for the great order and despatch with which the landing of the troops was effected under Brigadier-General Havelock, C.B. The highest spirit prevailed; and had the large Persian army only waited our approach out of the range of the ships' guns, I feel confident that it would have received a lasting lesson.'

The advance was made in lines of columns, the 23rd being in echelon on the left flank, with the front covered by skirmishers. When the enemy fled, the 23rd were ordered to capture and occupy the entrenched camp, which they successfully accomplished.

The enemy now retreated to Awhaz, one hundred miles up the Karun. But Outram followed close on their heels with his flotilla, and again completely defeated them; they fled upstream, abandoning huge supplies of grain and ammunition, and shortly after, news arrived that peace had been signed at Paris, Persia surrendering all claims to Herat and engaging not to interfere in Afghanistan. The 23rd, however, did not immediately return to India. They stayed behind as part of the force left to occupy Bushire and Kharak in order to enforce the terms of the treaty, and it was not until February,

1858, that they found themselves back in Bombay. For his services in Persia, Major Travers, the Commandant, received the thanks of the government, and on the return of the Division to India, the whole force received the cordial thanks of the Governor-General-in-Council 'for their orderly conduct and their steady discipline during the occupation of Bushire'. As a special mark of approbation, the Regiment was authorized to bear the word 'Persia' on its colours and appointments.

Meanwhile, stirring events were taking place in India. In May, 1857, the Mutiny had broken out, and Outram had been recalled with the brief comment, 'we want all our best men here'. The 23rd heard with admiration of their old adjutant's brilliant exploits in the relief of Lucknow, and their chagrin must have been intense when they arrived too late to participate in the most glorious exploit in the annals of the Bombay Army—Sir Hugh Rose's brilliant campaign in Central India, which culminated in the fall of the great fortresses of Jhansi and Gwalior, and the annihilation of the rebel forces, in which their sister regiment, the 25th, greatly distinguished themselves. But although when the 23rd once more landed in Bombay, the back of the Mutiny was broken, the danger was by no means over. The conflagration had been got under, but the glowing embers were still scattered far and wide, and might at any moment break into fresh flames. Many of the rebel leaders had been killed, or had taken sanctuary in the jungle of Nepal, but the most able of them all, the infamous Tantia Topi, had been headed off, and was being pitilessly hunted by flying columns. After receiving a rough handling at Jowra Alipore on June 22nd, he had fled to Rajputana, where he found a number of sympathizers, and was enabled to re-equip his army. There was now considerable apprehension that he would cross the Nerbudda and enter the Bombay Presidency. Were this to happen, the authorities, with very good reason, were apprehensive of a widespread rising, not only in the Poona and Satara districts, which were full of sympathizers with the Peshwa, but also in the turbulent Southern Maratha Country, which had never been properly subdued and had given a great deal of trouble on the outbreak of the mutiny. When it is remembered that Tantia Topi had a force of about nine thousand well-mounted men, it will be seen that the danger was considerable. It was particularly feared that he would attack Indore, where it was known that Holkar's troops were ready to join him if they thought they could do so with impunity.

Dispositions were therefore made to guard the line of the Vindhya range, in order that the rebels might not penetrate southward through an unguarded pass. For this purpose, Brigadier Hill, with the Hyderabad Contingent, was deputed to watch Asirgarh and Melghat, while further west, Sir Hugh Rose was in charge of a force guarding Khandesh. The 23rd formed part of the latter. From May to November, their head-quarters were at their old station of Malegaon, and the Regiment was constantly employed in sending detachments to various threatened points and to quell local disturbances in the Satpura hills. On November 8th an alarm was raised that Tantia Topi was about to attack the important town of Burhanpur, which commands the Grand Trunk Road, and thus interrupt communications between Bombay and Central India. To guard against this contingency, the Regiment was moved post-haste to the spot, and took up a position on the east of the city, where it remained for two nights in hourly expectation of being attacked. At one time the insurgents were within thirty miles of them, but it was not Tantia Topi's policy to assault entrenched positions, and he quickly decamped, to reappear at Chopra on the banks of the Tapti, which he appeared to be about to cross. In order to forestall this, the 23rd made a forced march of fifty-eight miles in thirty-eight hours, for which they received the thanks of Government and a special gratuity of three months' *batta*. Finding himself again headed off, Tantia Topi now threatened to make for the rich city of Baroda, where he had many sympathizers, and this necessitated fresh dispositions and more forced marches for the troops concerned. It is impossible to follow in detail the operations which ensued: detachments and flying columns were posted at every point of vantage, and were constantly being despatched in various directions, the head-quarters of the Regiment, from January 9th, 1859, being at Mhow. Owing to their constant alertness, Tantia Topi was at length compelled to give up his project of breaking southward, and made off at full speed for Northern Rajputana, but here again he was headed off. Driven out of North Rajputana, he next made for the dense jungles of the Banswara district, and here on Christmas Day 1858, he came into collision with a force under Major Rocke of the 72nd Highlanders, which contained a detachment, 250 strong, of the 23rd under Captain Waddington. Major Rocke after a sharp engagement prevented him from escaping through the pass at Pratapgarh into Malwa, and once more he broke northwards, trying in vain to shake off his relentless pursuers, who gave him no rest night or day. And now the hunt was up. Doubling hither and thither like a hare, Tantia Topi was at last tracked down in the pathless forests near Sironj, where he was betrayed by a confederate. On April 7th the arch-traitor expiated his crimes on the scaffold at Sipri, dying with a dignity worthy of a better cause. The Mutiny was now over, and the Regiment, having collected its scattered detachments at Mhow, was entrained for Bombay, which it reached on May 1st. The pursuit of Tantia Topi, in which it took so honourable a part, has no parallel in military history, save in the concluding phases of the Boer War.

AUTHORITIES

Regimental Records.

James Outram: a Biography, by Major-General Sir F. J. Goldsmid.

Kaye and Malleson's *History of the Indian Mutiny*.

A History of the British Army, book xiii, by the Hon. J. Fortescue.

REGIMENTAL LIFE, 1862-80

Regimental Silver, 123rd Outram's Rifles

[*face p.* 63

Chapter V

(1862-1880)

REGIMENTAL LIFE

In 1862 the Regiment was ordered to Karachi, and while there, it received the sorrowful news of the death of Sir James Outram, who, broken in health by the hardships of the military campaign, passed away at Pau in the South of France on March 11th, 1863, at the early age of sixty. He was laid to rest, among England's undying dead, in Westminster Abbey. Like many another Englishman, he had gone home to die, having given his life to the service of India. We have already referred to his early work among the Bhils and in Kolhapur, but a brief sketch of this great soldier's later career may perhaps not be inappropriate at this juncture. His first taste of active service on a large scale was during the operations in Afghanistan and Sind between 1838 and 1843, when he accompanied Sir John Keane as a member of his staff, and was the hero of many extraordinary exploits. At Ghazni, he captured a banner from the enemy. On another occasion, with a handful of British Cavalry numbering about one hundred, he went in pursuit of Dost Mohammad, chasing him over the Haji-Khak pass, 12,000 feet high, and then over the still higher pass of the Shutar Gardan, not relinquishing the pursuit until he reached Bamian and found that his prey had crossed the Oxus. At another time he volunteered to carry a dispatch for General Willshire from Kelat to Karachi by the direct route running to the Sonmiani Bandar on the coast and at the same time to reconnoitre the country with a view to its practicability for the passage of troops. On this occasion he literally took his life in his hands: the distance was 355 miles and the way lay through wild, unexplored country, and fanatical tribes who would have murdered him without hesitation. But Outram emerged safely: he had disguised himself as a Pir and for two days had subsisted on dates and water! In 1841 he negotiated a successful treaty with the Mir of Hyderabad, and was bitterly opposed to Napier's policy of the annexation of Sind. When the outbreak of hostilities became inevitable, he clung to his post at the Residency at Hyderabad until the last moment, and finally evacuated it with trifling loss in the face of a force of 8,000 Baluchis. With his usual nobility he refused to touch his share of the prize-money, and although he was a poor man, he devoted the entire sum, amounting to £3,000, to army charities. Always a champion of the oppressed, Outram was continually getting himself into trouble by his advocacy of those whom he thought to be harshly treated. In 1851, he incurred the displeasure of the Government of Bombay by his fearless, but indiscreet, exposure of *Khatpat* in Baroda State, where he was Resident. Of his campaign in Persia, we have already spoken; it was, however, completely eclipsed by the operations for the relief of Lucknow. It was characteristic of the man that, on his arrival at Cawnpore, he chivalrously refused to deprive Havelock of the honour of capturing Lucknow, and offered to serve under him in his civil capacity, with the Volunteer Cavalry. After cutting his way into the Residency under a tempest of fire, he held it until relieved by Sir Colin Campbell in November 1857.[1] His evacuation of the huge body of women and children from the Residency, over 1,599 souls, in the face of a watchful enemy, without a single casualty, was a masterpiece of organization. But his greatest achievement was to follow. After the relief, Sir Colin Campbell fell back upon Allahabad, leaving Outram with 5,000 men and 25 guns, to hold the Alam Bagh until he returned. Outram was at once assailed by the Moulvi Ahmed Shah, with the pick of the rebel army, consisting of 120,000 men with 130 guns. From December, until Sir Colin's re-appearance in March, he held them at bay, beating off attack after attack, a feat which has, not unjustly, been compared to Wellington's defence of the lines of Torres Vedras. These are only a few of the leading events of the life of this fearless soldier and great leader. But

[1] The Volunteer Cavalry unanimously voted him the V.C., but he characteristically cancelled the election on the grounds that he was ineligible as the general under whom they served! Yet the V.C. was the one reward in the world which he coveted.

great as he was as a leader, he was an even greater man. He was one of the first officers to take a genuine interest in the welfare of his men, whose health and comfort he carefully studied. He was a true lover of India and her people, and was as fearless in the exposure of what he considered to be wrong doing or injustice as he was in the field, utterly regardless of what it might cost him. The almost quixotic chivalry of his nature, his gentleness to the fallen and oppressed, his stainless character, and his fearless courage in the face of danger belong not to our days, but to the Middle Ages. Truly he was 'a very gentle, perfect knight', and it was his bitterest opponent, Sir Charles Napier, who publicly saluted him as 'The Bayard of India, *sans peur et sans reproche*'. These words are engraved on the marble slab which covers his grave, and never was a title more justly earned.

His life was given to India:
In early Manhood he reclaimed Wild Races by Winning Their Hearts:
Ghazni, Khelat, the Indian Caucasus, witnessed the daring Deeds of His Prime:
Persia brought to Sue for Peace, Lucknow Relieved, Defended and Recovered
Were Fields of His Later Glories.
Many Wise Rulers, many Valiant Captains hath this Country sent Hither;
But never any Loved as this Man was, by those whom they Governed or Sent to Battle!
Faithful Servant of England
Large Minded and Kindly Ruler of Her Subjects,
Doing nought through Vain-glory, but ever esteeming others better than Himself;
Valiant, Incorrupt, Self-denying, Magnanimous;
In all the True Knight!

So runs the noble epitaph inscribed upon his monument on the Calcutta Esplanade, and never was praise more richly earned. The name of Outram will always be held in reverence by the regiment which has the honour to bear it, and his character and achievements will be an undying example to the generations of officers who serve in it.

From January 1st, 1864, an important reorganization of the Indian Army came into force. The three Presidency Armies were reorganized on the so-called 'irregular' system, under which the British officers acted only as field and regimental Staff Officers. Henceforth, each battalion was to have as its establishment of officers a Commandant, Senior and Junior Wing Commandants and Adjutant, a Quarter-Master, and a Doing Duty Officer. Under the new organization, the following officers were posted to the Regiment :—

Commandant	Lieut.-Col. S. J. Whitehill.
Senior Wing Commandant ..	Major J. Peyton.
Junior Wing Commandant ..	Captain E. Waddington.
Adjutant	Lieutenant E. R. Ross.
Quartermaster	Lieutenant J. Gatacre.
Doing Duty Officer	Lieutenant F. Paul.

Under this system, it will be seen, the command of companies was placed in the hands of Indian Officers, only the Commanding Officer, the Wing Commandants and the Regimental Staff being British. The system, which had been found to work well in the Mutiny, had the merit of economy, and at the same time developed the initiative and sense of responsibility of the Indian Officers. That the Regiment was keeping up its high standard of efficiency is shown by the inspection reports of the time. In 1864 Major-General Honnor, commanding the Sind Brigade, testifies to his satisfaction at its 'standing on parade, its correctness in manœuvring, its cleanliness and general conduct', and in the following year, Sir William Mansfield, the Commander-in-Chief, in a farewell order to the Karachi Brigade referred to 'the most favourable reports received by the Regiment during the last five years', adding that 'it afforded him great pleasure to be able, by personal observation, to say that these reports were fully justified'.

In October 1865 the 23rd was ordered on Field Service in Okhamandal. Ever since the British had taken over Kathiawar, endless trouble had been experienced with the turbulent tribe of the Vaghirs. Their head-quarters were in the little peninsula of Okhamandal, in the north-west corner of

Kathiawar, and many of them made their living by piracy, and by plundering peaceful traders, both on land and sea. Expeditions sent against them in 1816 and 1820 had effected little. In 1862, several Vaghirs who had been imprisoned at Rewa Kantha broke jail and started a series of fresh disturbances; whole villages were burnt and devastated, and peaceful inhabitants murdered wholesale by these bloodthirsty outlaws. The 23rd reached Dwarka on October 9th, and detachments were at once despatched to Poshitra and Karanga. On November 9th, a portion of the Poshitra detachment, consisting of forty-two men under Subedar Shankar Pathak, had a sharp encounter with 150 outlaws under a noted leader named Ranaji, who were posted in a strong position in dense jungle. Subedar Shankar Pathak's report on the engagement runs as follows:—

'At about 1 p.m. yesterday, I received information from the Thanadar at Gopee, that a body of robbers were in the Dabla Jungle, near Gopee, and to come without delay. I, therefore, started and came upon them about 2 o'clock: there were about 150 as far as I could make out. I had with me forty-two Non-commissioned, rank and file. The Grassia[1] of this place went with me (by name Kessree Singh). The outlaws, on our surprising them, rose up and appeared to be making off. I ordered the men to fire into them; a few shots were returned, but before we could reload, they rushed upon us with their drawn swords. I ordered the men to fix bayonets and charge, driving them back into the thick jungle: we were in skirmishing order at the time. One of the outlaws, whose name I have since ascertained to be Ranaji, and who was one of their leaders, came at me, and I having only the ordinary sword, he cut me across the face, and wounded me severely in the wrist, but fortunately Naik Surnam Singh, No. 4 Company, and Colour-Havildar Khooshal Singh, No. 4 Company, came to my rescue and bayoneted him, otherwise I should have lost my life. I was then carried to the rear, being disabled. We fought for two hours and a half, the rebels being driven back into the thick jungle, where I did not think it prudent to pursue them on account of the weakness of my force. Mooloo, the

[1] Petty chief or landholder.

chief of the outlaws, was present, but never came out of the thick jungle.

'We killed three of the outlaws, viz., Ranaji, Mooloo Vaghir (Mooloo Sajun's son), and a Vaghir of Burdia.

'I should fancy that some twenty or twenty-five were wounded, and I saw five or six fall on the first volley, who were carried away.

'Our loss in killed is as follows :—

1. Naik Roopram Pandy, No. 1 Company.
2. Private Govind Lard, No. 1 Company.
3. Private Luximon Bhoojbul, No. 1 Company.

Wounded.

1. Subhedar Sunkur Patuck, No. 3 Company, face and wrists, sword cuts.
2. Private Durga Hulwae, No. 3 Company, wrist bullet.
3. Private Dajee Goreykur, No. 1 Company, hand, sword-cut, spent ball in knee.'

For his conspicuous gallantry in action on this occasion, Subedar Shankar Pathak was awarded the Third Class of the Order of Merit. Unfortunately, this was the only engagement which the Regiment had with the outlaws. The villagers were so terrified that they did not dare to give any information as to their whereabouts, and the rebellion went on until December 1867, when the Vaghirs were defeated and practically exterminated in a pitched battle against a force under Major Reynolds of the 17th Native Infantry. The desperate nature of the fighting is shown by the fact that Major Reynolds was dangerously wounded and two British officers were killed in the engagement. It was not until November 1868 that the 23rd was able to return to Bombay, leaving two companies for detachment duty at Dwarka.

The Regiment was not destined to see any further active service for the next twelve years, but this long period of peace was marked by several interesting events. In 1867, volunteers were called for to join the 18th Infantry, which was under orders for Abyssinia. Twenty-four men of the 23rd were selected, and five for the Land Transport Corps. On March 12th, 1870, the Regiment was congratulated for the great alacrity with which it turned out for a fire alarm at Yeravda. On October 19th of the same year, an interesting ceremony took place, when Lady Spencer presented the Regiment with new Colours. On doing so, Lady Spencer addressed the regiment in the following words :—

'Colonel Whitehill, I feel that you have done me great honour in requesting me to present new colours to the distinguished Regiment under your command. On looking over the records of the 23rd Regiment of Light Infantry, I observe that they, with the rest of the Peshwa's Brigade, tendered allegiance to the British Government previous to the Battle of Kirkee, and in acknowledgment of their services on that occasion they were subsequently allowed to adopt the word *Kirkee* as an honorary badge, also that they received their first colours at Poona from the hands of Lady Colville.

'The career of the 23rd Regiment Native Light Infantry, has since that time been one of uninterrupted success.

'Colonel Whitehill, I have now only to express my heartfelt wishes, that success may attend the 23rd Regiment whenever they may be called upon to act in defence of their Queen and Country, and that their past achievements may, if possible, be surpassed by the future, when fighting under the colours which it is this day my pleasing duty to present.'

Colonel Whitehill in reply said :—

'Lady Spencer, allow me in the name of the officers and men of the 23rd Regiment to thank you most sincerely for the great honour you have done us in presenting to us our new colours.

'It is a ceremony of the greatest interest and importance at all times to a soldier, but I cannot but feel that on the present occasion it is doubly so, receiving our colours from the hands of the wife of so distinguished an officer as His Excellency the Commander-in-Chief. Your kind mention of the services of the Regiment I fully prize, and I cannot but feel proud of being Commandant of this Regiment, all ranks of which will, I feel assured, as I do myself, prize your kindness, when I make known all you have said to them before I dismiss them, which

I shall do. Allow me to repeat our sincere thanks for your great kindness.'

At the conclusion of the parade, the Commander-in-Chief expressed his high approval of the appearance and movements of the Regiment. The old Colours were duly deposited in All Saints' Church, Kirkee, where they hang to-day. From 1871 to 1880, the 23rd was stationed in various cantonments—Nasirabad, Mhow, Indore and Ahmednagar. On March 9th, 1876, the Indore detachment was present at Chowral Station on the occasion of the State visit of H.R.H. the Prince of Wales. Forty-five years later, by a curious coincidence, it furnished the Guard of Honour to his grandson on a similar occasion. In 1877 they were re-armed with Snider rifles, and in the following year, an incident took place which showed the fine spirit animating all ranks. In consequence of a threatened war with Russia, Lord Beaconsfield took the unprecedented step of sending seven thousand sepoys to occupy Malta. On volunteers being called for in April to join the 9th Bombay Infantry, ordered on Field Service, nearly the whole Regiment presented itself, but only seven men were required, and these accompanied the force which went to Malta. As soon as it became known that there was a prospect of service in the field, the Indian Officers of the Regiment waited on the Commanding Officer—Colonel Bates—and the men paraded outside the Colonel's bungalow, and with the greatest enthusiasm volunteered for service in any part of the world. The Commanding Officer was met with loud applause, and when he harangued the men, and told them he would make known their request to the Commander-in-Chief, his remarks were received with loud cheers which were continued as the men dispersed to the Lines. His Excellency the Commander-in-Chief, Sir Charles Stavely, K.C.B., expressed himself in reply much pleased with the very excellent spirit existing in the Regiment.

AUTHORITIES

Regimental Records. Captain W. A. M. Wilson's *Historical Records of the 23rd Regiment*.

THE SECOND AFGHAN WAR

CHAPTER VI

(1880-1886)

THE SECOND AFGHAN WAR

IN 1878 our relations with Afghanistan once more became strained. The Amir, Sher Ali, was offended at our occupation of Quetta and received a Russian Embassy at Kabul, while a counter-embassy, despatched by the Government of India, was turned back at Ali Masjid, the Afghan fortress at the head of the Khyber Pass. As no reply was received to an ultimatum, Afghanistan was invaded by three columns led by General Samuel Browne, General Roberts and General Stewart. Little opposition was encountered: a treaty was signed at Gandamak, and our envoy Sir Louis Cavagnari proceeded to Kabul. Suddenly, however, the world was horrified at the news of the treacherous murder of Cavagnari and his companions. War was immediately resumed. General Stewart occupied Kandahar and General Roberts entered Kabul. Arrangements were being made to hand over the country to Abdurrahman, a nephew of Sher Ali, the late Amir, when a serious disaster occurred. Ayub Khan, a son of Sher Ali, who had established himself at Herat, was threatening Ghazni and Kandahar. General Burrow, with a brigade of Bombay troops, was sent from Kandahar to watch the fords over the Helmand river; unfortunately he advanced to the stream and allowed himself to be caught in an impossible position at Maiwand by the whole of the Ayub Khan's army, numbering 20,000 men. His Afghan allies had already deserted him, and he was completely defeated, losing over 1,000 out of 2,500 men. The remnant fell back in disorder on Kandahar, which was soon closely invested, while Quetta was completely isolated (July 27th, 1880).

On February 7th, 1880, the 23rd received orders while at Ahmednagar to proceed on Active Service, and to join the reserve division ordered to concentrate in Sind. They sailed from Bombay for Karachi on March 10th. Colonel J. Harpur was Commandant at the time, and the Battalion, with seven European officers and 621 other ranks, reached Sukkur on the 17th. They there formed part of the Reserve Division, whose duty it was to keep open the line of communications between Quetta and the Indus Valley. The Head-quarters and Left Wing of the battalion under Colonel Harpur proceeded to Jacobabad on April 5th, and arrived at Sibi on May 5th. It was the height of the hot weather; the troops were under canvas, and the thermometer at midday registered 120° in the shade. At the end of the month, the Left Wing relieved the 1st Baluchis, and took over from them the Sibi outpost lines. During the whole period that head-quarters were at Sibi, from May 5th to August 21st, 1880, the Regiment was broken up into small detachments, and employed on outpost duties and in supplying convoy and escort guards. This work was rendered terribly trying and harassing owing to the extreme heat; many deaths occurred from heat-apoplexy, including that of the M.O., Surgeon-Major Simpson, who was buried in Sibi cemetery on August 31st. Meanwhile the Right Wing, under the command of Major Gatacre, with Lieutenant Tobin, crossed the frontier on April 2nd, and proceeded by train to Sibi, where they proceeded to occupy the posts on the Sibi-Harnai route as far as Spintangi. From the very first, a great deal of trouble was experienced from the turbulent Marris and Pathans, who constantly raided the lines of communication. On June 7th Major Gatacre left Lieutenant Tobin in charge of the Spintangi post, and went off to chastise some raiders in the Sangam Valley, who had looted the Commissariat Stores, and carried off Government transport animals grazing there. On August 4th Lieutenant Tobin, with part of the Spintangi garrison, attacked and dispersed another body of raiders who had assembled at Sinari to waylay and plunder the convoy from Harnai. In a personal encounter with a Pathan, Lieutenant Tobin had his horse wounded and received severe cuts on the neck and sword-arm, but succeeded in killing his adversary. The raiders were driven off with a loss of ten killed. Two days later Lieutenant Tobin was engaged in a more serious affair. After the disaster at Maiwand, it had been decided to withdraw all the troops from the outposts on the Harnai route back to Sibi ; the work on the projected light railway was stopped, and all treasures, stores and materials, together with about two thousand railway labourers and a large number of bullock carts, had to be escorted thither, On reaching Kachali, it was found that the defile was held by a large body of Marris. A Council of War was held, consisting of Dr. Duke, the Political Officer, Major Peters, R.E., of the Railway Staff, and Captain Bowles of the 15th Foot, the Transport Officer. Lieutenant Tobin was too ill, owing to his wound, to take any part. It was decided that the only thing to do was to push on to Sibi at all costs.

News of this was soon spread, and the coolies, to the number of about two thousand, seized with panic, took the bullocks out of the carts containing the wounded and treasure, and set off in a mob down the pass. Major Peters and Captain Bowles both decided that they were for the time non-combatant, and Lieutenant Tobin, severely wounded as he was, was left to do the best he could. He sent off the few sowars of the Sind Horse he had with him to head the fugitives, while he quickly collected his own men, who were cooking when the stampede commenced. The Marris by this time had come down, and were busy murdering the helpless coolies, who broke back, almost overwhelming the small party of sepoys in their rush. Lieutenant Tobin managed to drive the Marris back, and then the retreat through the pass commenced. What took place can be imagined—a narrow pass, a sweltering hot day, a wounded officer in command, a handful of sepoys, and some two thousand unarmed, undisciplined coolies, beset on all sides by a horde of tribesmen ! The party reached Sibi on the 8th, the casualties being one Naik, six Privates, one Lascar, one Bhisti, and two Kahars killed, and six privates wounded.

Lieutenant Tobin received the special thanks of H.E. the Commander-in-Chief and the Government of India 'for his excellent behaviour under the very difficult circumstances in which he was placed on that occasion'. It was remarked that 'in fact, he seems to have done everything that man could do'.

Immediately on receipt of the news of the defeat at Maiwand and the investment of Kandahar, arrangements were made to despatch simultaneously two strong divisions to the relief of the beleagured force, one from Quetta under General Phayre, and the other from Kabul under Sir Frederick Roberts. Accordingly, the 23rd received orders to proceed to Quetta. They left Sibi on August 21st, and marching by night in order to avoid the heat, crossed the Bolan Pass and reached Quetta on the 29th. After a day's halt to rest and reorganize, they continued their advance, in charge of a convoy of three hundred carts, laden with food supplies urgently needed for the front. On the road an incident occurred which illustrated the fine spirit animating all ranks. The drivers suddenly took fright and deserted in a body, and could not be replaced. In this dilemma, Captain Watling, D.A.Q.M.G., arranged with Colonel Harpur that three hundred sepoys, or about half the regiment, should act as drivers. This they readily did, and after incessant labour—the draught cattle being so weak and ill-conditioned that they could scarcely draw their loads—they brought this important convoy safely to its destination. For this important service they received the well-earned approbation of General Phayre who in his Despatch of October 16th, 1880, drew special notice to the instance, as showing ' the zealous spirit actuating the troops with the rear brigade of the Division '. On September 6th, the Regiment reached Killa-Abdulla, and there it became part of the 3rd brigade, the other regiments being the 63rd Foot and the 9th Bombay Native Infantry. They arrived at Kandahar on September 23rd. It was unfortunate that General Phayre's force, after toiling through Sind and the Bolan Pass at the hottest time of the year, was delayed, in spite of every exertion to overcome these obstacles, not only by formidable transport difficulties, but also by the hostile attitude of the tribes on the lines of communication. These had to be kept open at all costs, as they were the sole means of supplying the Kandahar garrison and the Kabul column. For these reasons the division arrived at Kandahar only to find that Sir Frederick Roberts had already relieved it on August 30th, defeating Ayub Khan's force with great slaughter soon after. There the Regiment remained until October 22nd, when it was told off for duty on the line of communications, and furnished outposts of two companies each at Abdurrahman, Mel Karez, Dubrai and Gatai.

On April 24th, 1881, on the evacuation of Kandahar by the British troops, the Regiment joined the last Brigade of General Hume's Division and marched for Quetta, which was reached on May 8th. They left Quetta on the 14th, arrived at and left Pir Chauki on the 21st, and arrived at Karachi on the 23rd. Major Gatacre with 120 men was left at Karachi, while the remainder embarked on the 24th on board the I.G.S. *Dalhousie*, arriving at Bombay on the 27th, and at Ahmedabad on the 28th. On June 4th the detachment from Karachi rejoined. From July 12th throughout the stay of the Regiment at Ahmedabad one company remained on detachment at Sadra.

On July 29th, 1881, the Regiment was authorized to wear upon its colours and appointments the words 'Afghanistan 1879-80'. On August 11th, 1883, at a parade held at Ahmedabad, Brigadier-General Carnegy presented the regiment with the medals for the Campaign, and addressed them as follows :—

'Colonel Gatacre, Officers and men of the 23rd Native Light Infantry, I can conceive no more agreeable duty that a General Officer can be called upon to perform than that of being the medium of distributing decorations by Her Most Gracious Majesty the Queen-Empress on her troops in recognition of their loyal and faithful services during times of trial and danger, and I thank you most sincerely for giving me this opportunity of presenting you with these medals which were so honourably won by you in the recent campaign in Afghanistan. Although circumstances prevented your sharing in any of the chief or what may be termed the general actions of the campaign, yet the duties devolving on the Regiment on the line of communications were carried out in such a manner as tended in no small degree to the ultimate success of the campaign, and gave evidence of that strict discipline which on more than one trying occasion proved that when broken up into small detachments, your officers and non-commissioned officers possessed the self-reliance and confidence, which inspired an equal confidence in the men they commanded.' General Carnegy recounted the history of the Regiment during the Afghan war, and alluded to the leading events in its records, and in conclusion said: 'Officers and men, I again repeat that I feel grateful to you for permitting me to distribute the Afghan medals to your distinguished regiment. And I trust that a recollection of the past services of the Corps and the noble names which are associated with its past deeds will stimulate you and your successors to even greater deeds for the honour of your Queen-Empress and country.'

There is little more to add about the Regiment during the brief period between its return to India in 1882 and its departure on Field Service in Burma in 1886. After the Mutiny the uniform of the Sepoys, which had become assimilated to that of British troops, causing intense suffering in campaigns like those in Sind and Central India in the hot weather, was much modified. The Kilmarnock cap was replaced by the *pagri*, and in 1882 the uniform was changed to a serge jacket and knickerbockers, with khaki for summer wear. Khaki appears actually to have been first worn by the battalion at Kandahar in 1880, when the red uniforms were left behind at Quetta. In the same year, the Regiment volunteered for service in Egypt, but, to the great disappointment of all concerned, it was not selected. On July 1st the establishment was fixed as under: one Commandant, one Second in Command, one Wing Commander, one Adjutant, one Quartermaster, three Wing Officers, one Medical Officer, 16 Native Officers, 40 Havildars, 16 Buglers and 760 rank and file. The British officers then with the regiment were as follows:—

Commandant	Colonel J. Harpur.
Second in Command ..	Major J. Gatacre.
Wing-Commander	Major F. T. Ebden.
Wing Officer	Captain E. C. Kellie.
Quartermaster	Lieutenant W. St. J. Richardson.
Adjutant	Lieutenant R. I. Scallon.
Officiating Wing Officers ..	Lieutenant J. W. C. Hutchinson.
	Lieutenant C. G. Norris.
	Lieutenant H. Parkin.
Medical Officer	Surgeon J. C. Lucas.
Major J. H. Watling (Seconded).	

The Regiment continued to obtain highly favourable reports from inspecting officers, who unanimously praised it for its turn-out on parade and steadiness under arms, and in 1885, while at Neemuch, it furnished a Guard of Honour for H.E. the Viceroy, Lord Dufferin, which was especially noticed for its smartness. For three years in succession, 1883-6, it was first in musketry in the Bombay Army, and the achievement of a squad at the Poona Rifle Meeting in 1884 was the subject of a special letter from Major-General Carnegy to Colonel Gatacre. Referring to the numerous prizes carried off by the squad, and especially by Sardar Bahadur Subedar Salvadore Gabriel, General Carnegy remarked that these results could not have been achieved without great care and attention to musketry on the part of the officers, and more especially the Adjutant, which had raised the corps to its 'proud position as the best-shooting Native Regiment in the Bombay Army'.

Appendix

RECORD OF SERVICES OF OFFICERS OF THE 23RD BOMBAY NATIVE LIGHT INFANTRY WHO SERVED IN THE AFGHAN CAMPAIGN, 1880-1

Colonel J. Harpur. Commanded the regiment throughout the period it was on active service: O.C. Troops at Sibi, May 5th-August 21st, 1880. Retired January 1st, 1885 with a Good Service Pension.

Major J. Gatacre. Served with the regiment, in command of the Right Wing, throughout the period it was on active service. In action against the Marris in the Sangam Valley; recovered looted property and destroyed their village. In command of Gulistan Fort September 5th-10th, 1880, and the outposts at Gatai and Dubrai, October 25th, 1880-April 26th, 1881. Promoted Lieut.-Colonel, June 13th, 1883.

Lieutenant W. St. J. Richardson. Transferred to 1st Belooch Native Infantry, May 1880.

Lieutenant R. I. Scallon. Served with 2nd Belooch Native Infantry, November 1879-September 1880. Present at the actions of Khan Khel, Kaj-Baj, and Kandahar, and accompanied Brigadier-General Daubeny's column to Maiwand.

CAPTAIN J. T. WATLING. D.A.Q.M.G., April 16th, 1878. Mentioned in despatches for his excellent staff work on the Kandahar-Nari and Bolan lines of communication. Promoted Major, November 26th, 1881. A.Q.M.G. on the establishment, October 31st, 1882.

LIEUTENANT F. J. TOBIN. Commanded the detachment engaged with the enemy on August 4th, 1880. Severely wounded. Commanded detachment of the 23rd Native Light Infantry and Bombay Cavalry during the withdrawal from Sibi, when he repulsed a large body of Marris. Received the thanks of the Government of India and C.-in-C. Struck off the strength, September 21st, 1882.[1]

SURGEON-MAJOR J. SIMPSON, M.D. Regimental M.O. since 1872, died of heatstroke and buried at Sibi, August 13th, 1880.

AUTHORITIES

Regimental Records. *The Afghan Campaigns of 1878-1880*, by S. H. Shadbolt (1882).

Forty-one Years in India, by Lord Roberts (1897).

Cambridge History of India, Vol. VI, chapters xxii, xxiii.

[1] Lieutenant (now Colonel) F. J. Tobin joined the 23rd N.L.I. as a Probationer in 1879. Owing to ill-health, he was sent back by the medical authorities to his British Regiment, the 86th Foot (Royal Irish Rifles), in 1882. He commanded them in the South African War, where he was awarded the D.S.O., and served in the Great War, 1914-18.

THE THIRD BURMESE WAR

Major-General Sir John Gatacre, C.B.

[face p. 83

Chapter VII

(1886-8)

THE THIRD BURMESE WAR

The word of a scout—a march by night,
A rush through the mist—a scattering fight,
A volley from cover—a corpse in the clearing,
The glimpse of a loin-cloth and heavy jade earring,
The flare of a village—the tally of slain,
And—the Boh was abroad 'on the raid' again!

Kipling, *The Ballad of Boh da Thone.*

In 1885, relations with Burma were daily becoming more and more strained. King Theebaw, a vain and cruel tyrant, cordially disliked the presence of the English in Lower Burma, and was especially hostile to the Burma Trading Company; on one occasion he imposed upon them the enormous fine of twenty-three lakhs of rupees, and ordered the arrest of the employees. Furthermore, he was constantly intriguing with the French, and Lord Dufferin learnt that he had concluded a treaty with the French Government, giving them exclusive consular and commercial privileges of an extensive character. French predominance at Mandalay would be an even greater danger to British India than Russian predominance at Kabul, and after prolonged protests on the part of the Government, which Theebaw treated with contempt, a force was despatched to Burma on October 21st. The first part of the campaign was a very trivial affair. Mandalay was occupied, King Theebaw surrendered and was banished to India, and Upper Burma annexed (January 1st, 1886).

But the real trouble began when the formal war was finished. The bands of dacoits, which had long been looked upon by young Burmans as a suitable outlet for their love of adventure, were reinforced by the disbanded soldiery of the late monarch. The whole countryside was in a blaze in a very short time. In addition, it must be remembered that the fighting took place in thick and almost impenetrable jungle, of which the enemy knew every inch, and the inhabitants of the villages were in thorough sympathy with their countrymen. The rainfall in Upper Burma is heavy, and during the wet season much of the country is a vast swamp, while malaria takes a heavy toll of all who are obliged to pass through it. The third Burmese war has often been called 'The Subaltern's War'. Pitched battles were impossible; the fighting was carried out by small columns, who were constantly on the move, attacking stockaded villages, escorting convoys from one fortified post to another, or covering working parties. A volley poured in without warning at close quarters by an invisible foe, who disappeared as quickly as he had come, was an almost daily occurrence, and nothing is more calculated to shake the *morale* of the troops employed in these operations. The Burman, though undisciplined, is a brave and determined fighter in his own country, and it finally took 30,000 troops five years' hard work to make an end of this guerilla campaign. These facts will give the reader some idea of the formidable nature of the operations which the 23rd were called upon to undertake.

The Regiment was at Neemuch when it was warned for Field Service on July 5th, 1886, and it left Bombay for Rangoon on the *Clive* on the 12th of the same month. Its strength was nine British officers, twelve Indian officers, and 606 other ranks. The following were the British officers serving with the Regiment at the time of embarkation :—

Commandant	Lieut.-Colonel J. Gatacre.[1]
Second in Command	Major W. C. Black.
Wing Commandant	Captain E. C. Kellie.
Adjutant	Lieutenant R. I. Scallon.[2]
Quartermaster	Lieutenant S. W. Lincoln.
Wing Officers	Lieutenant W. A. M. Wilson.
	Lieutenant W. L. Conran.
	Lieutenant W. L. Warner.
Surgeon	C. B. Maitland.

[1] Colonel, June 13th, 1887.

[2] Captain, February 12th, 1887.

UPPER BURMA

96°
98°
Chindwin R.
SHWEBO
ALON
MONYWA
LOWER CHINDWIN
22°
MANDALAY
MAYMYO
SAGAING
KYAUKSE
MONGKUNG
MYINGYAN
MYITTHA
PAUK
PAKOKKU
MAHLAING
SEIKPYU
MEIKTILA
HLAINGDET
TAUNGGYI
MONG PAWN
YAMETHIN
NATMAUK
MYOTHIT
MINBU
20°
Scale 1" - 32 Miles
Fr. Rangoon

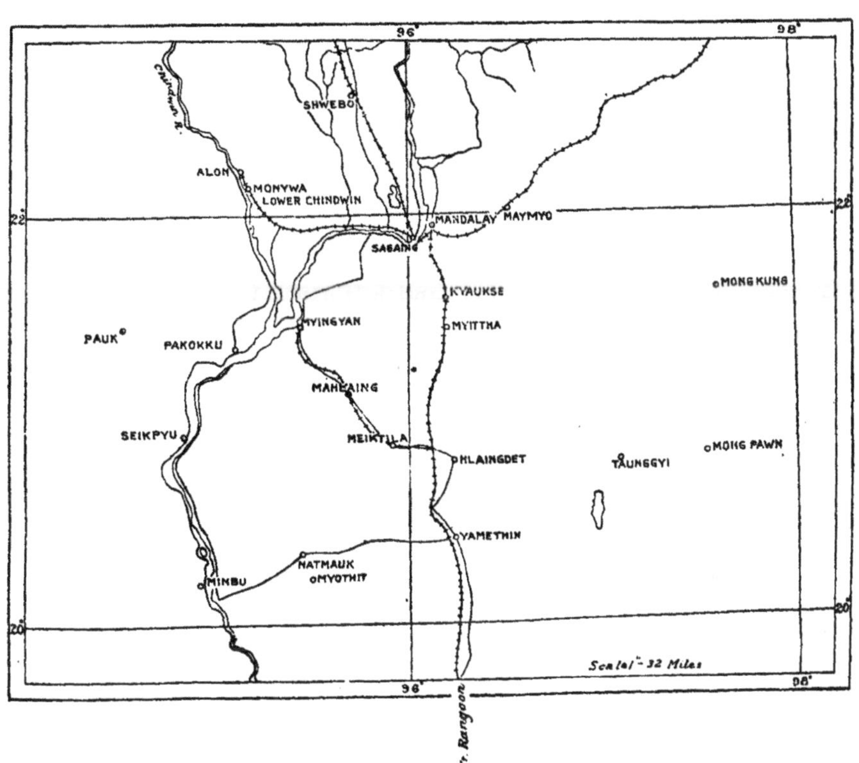

In the course of the campaign a number of changes necessarily occurred. Major Black was transferred to the Staff, and his place was taken by Major F. T. Ebden. Lieut. W. S. Delamain rejoined the Regiment on September 9th, 1887.

After a rough passage through heavy monsoon seas, the Regiment reached Rangoon on the 22nd, and transhipped to the *Eden* of the Irrawady Flotilla.

The first station of the head-quarters of the 23rd Bombay Light Infantry was at Myingyan in Upper Burma, at the junction of the Chindwin and Irrawady, where it remained from August 5th, 1886 to April 3rd, 1889. The detachments furnished by the Regiment during this time were Pyilingyaw, Senbyagyoun, Salen, Natogyi, Yetagyo, Sametkyon, Mehtyila, Pakhangyi and Shimadoung, and latterly a large detachment for the Pakoku-Kanle column. The work done by the men at head-quarters was hard on account of the large number required to escort ration and other parties, as well as to supply guards to river steamers. Small parties were also continually out on ' daurs '. From April 4th, 1887 to April 27th, 1888 the head-quarters were stationed at Pakoku. During this period the detached posts garrisoned were Pakhangyi, Shimadoung, Myaing, Kanle, Tebya, Lingadaw, Pyinchoung, Pauk and Poungloung, and for a short time there were detachments at Alon and Mingin up the Chindwin River. Later on, in the cold weather of 1887-88, two large columns were sent out—the one to operate in the Kyaw Valley and the other in the Yaw Valley. At the same time Colonel Heyland, 1st Bombay Lancers, was operating in the Pakoku District against Boh Nga Kwe, and several small posts were garrisoned to support him. The work of rationing all these posts and parties fell for the most part on the men at head-quarters, and in consequence of the small number left there, the escort duty was very heavy.

The Pyilingyaw detachment consisted of Lieutenant W. A. M. Wilson and 105 rank and file. It was stationed at a village of that name on the Mon Creek in the Mimbu District and remained there from the beginning of August to the end of October 1886. The country round was in a very disturbed state owing to the murder of Mr. Phayre, Deputy Commissioner, and the withdrawal of the troops composing the garrison of Ngapeh. The two principal dacoits in this neighbourhood were the well-known Boh Shwe and the Phoongyi Boh Oktama. The country was difficult to operate in on account of the thick jungle on one side and paddy cultivation on the other, the heavy rainfall and the numerous large nullahs and creeks intersecting the country. The men, though just arrived in the country, adapted themselves to the work with great keenness. There were numerous and at one time almost daily skirmishes with the dacoits. The most notable engagement with the enemy was on September 3rd, 1886 at Phayagon, where the Burmans had taken up a very strong position inside the enclosure of a pagoda with the intention of ambuscading a party of Punjab Police. They were surprised by a party of the 23rd Bombay Light Infantry, and after a sharp fight were driven out with the bayonet. The civil authorities reported that on this occasion there were thirty-two killed and about seventy wounded. For services on this day the Order of Merit, Third Class, was conferred on Subedar Salvadore Gabriel and seven men of the detachment, and Major-General White expressed his approbation of the operations. Boh Oktama was surprised in the village of Tokoko on the 9th September, 1888. The party of the 23rd Bombay Light Infantry, numbering sixty, had to advance through paddy fields under fire, but drove out the dacoits, who were reported to number three hundred, killing nineteen and wounding a large number. Extensive patrols through the district were carried out, although the rain was very heavy and the country under water. Major Parrot, Deputy Commissioner at Mimbu, repeatedly expressed his high opinion of the work done by this detachment in keeping quiet a large tract of country between the Mon and Ma creeks. Major Carre, R.A., and Lieut.-Colonel Rolland, M.S.C., commanding the neighbouring post of Segu, also expressed thorough approval of the work done.

The Senbyagyoun detachment consisted of a jemadar and fifty non-commissioned officers and men. The principal object of the detachment was to keep open the line of communication between Salen and the river, which had constantly been interrupted by dacoits, and in this it was entirely successful. The party was detached from August 3rd, to the end of October 1886.

The Salen detachment consisted of Major W. Black, Lieutenant L. Warner and one hundred Indian ranks, and was away from 3rd August to the end of October 1886. It was landed while the regiment was proceeding up the river, to reinforce the garrison of Salen after the fierce attack on that post in July 1886. The relief of Salen had the effect of breaking the dacoits up into small bands, which were dispersed all over the surrounding district. Although this detachment was not fortunate enough to have any engagement of importance with the enemy, its constant patrols had the effect of driving the dacoits from the district and prevented them again gathering and making any head.

The detachment at Natogyi in the Myingyan District, consisting of Captain E. C. Kellie and 102 Indian ranks, remained there from August 5th to the beginning of November 1886. The country round was to a large extent under paddy cultivation with the very roughest of tracks to serve as roads. The principal engagement in which a portion of this detachment took part was the surprise of Nabuddah on September 2nd, 1886, when twenty-five dacoits were killed and a large number wounded. The whole of the country in the vicinity was regularly patrolled, and in the numerous skirmishes which took place many dacoits were killed and wounded.

In November 1886 the Natogyi detachment was moved twenty-two miles further on to Pyinzimyo, and remained there till March 31st, 1887. Here much excellent work was done. A small party of Mounted Infantry was instituted, which, mounted on ponies captured from dacoits, was able to make more extensive patrols. Numerous night marches were made to surprise villages which were known to harbour dacoits, and with successful results, as at Nahimyit, where eight dacoits were killed and a large quantity of arms taken, and at Myitshoo, where the enemy lost twelve men.

In the middle of August 1886 a detachment consisting of two British officers—Lieutenant Lincoln and Lieutenant Conran—and seventy-five Indian ranks, was sent to Yetagyo on the Chindwin. Parties from this post had several successful engagements with dacoits, such as at Maoo, where a party of dacoits was caught crossing the Chindwin and lost ten men. In reference to this expedition, a letter was received from the Assistant Adjutant-General, Upper Burma Field Force, saying that the General Officer Commanding considered the operations by Lieutenant Lincoln were ' well conducted '.

From October 31st, 1886 to March 14th, 1887 a detachment of one Subedar and fifty other ranks was stationed at Sametkyon, a post on the Irrawady, chiefly to guard the village, which was a halting place for steamers. Patrols were sent out on both sides of the river.

A detachment, consisting of one European officer—Major Ebden—and 103 Indian ranks, was posted at Meiktila from October 30th, 1886 to January 22nd, 1887. This had originally been a very disturbed district, but, owing to the presence of the troops, soon became comparatively quiet, and, although parties of the detachment were constantly out, no large body of dacoits was encountered.

A detachment of varying strength was stationed at Pakhangyi from November 3rd, 1886 to May 17th, 1887. The party from Yetagyo first took up the post strengthened by about fifty men from regimental head-quarters, making the total of Indian ranks about 125 with two British officers. The situation three miles from the Chindwin was most unhealthy, as during the rains the intervening country was under water, and when drying up was the cause of much sickness. At one time as many as seventy men of the detachment were suffering from fever, but the duties were carried on without intermission. At the time of first occupation the large and important town of Pakhangyi had been burned by the dacoits and the surrounding villages had been deserted by their inhabitants. The establishment of this post did much good by restoring confidence and inducing the people to return to their homes. The dacoits, who for some time had been doing much harm in the villages on the Chindwin between Yetagyo and Pakoku, retreated inland towards Kanle. Extensive and distant patrols were sent out from this post, and a large quantity of arms was collected. One expedition from this post to the islands at the mouth of the Chindwin drove the dacoits from the villages of

Soolaygon, Oneinbok, and Pyema, which they held successively, and finally forced them to cross the Chindwin. On this occasion the dacoits lost a considerable number in killed and wounded.

Shimadoung was held by a detachment of the Regiment from December 5th, 1886 to the end of April 1888. This post, situated on top of a hill about 1,750 feet high, was taken up as a signalling station to enable the posts and columns of the Pakhangyi, Pakoku, and other districts to communicate with the brigade head-quarters at Myingyan. The chief hardship which this detachment had to bear was scarcity of water. When the post was first occupied, a small supply of water was found in a few cisterns cut out of the rock on the hillside. This was speedily exhausted, and the men of the detachment had to go daily to the bottom of the hill to procure their water, at first from a tank one mile from the foot of the hill, but in the hot weather from a tank two miles distant. The labour of carrying the water was excessive, as the hill was so steep and the roadway consisted the whole way of rough and broken steps. On one occasion two men of the detachment were killed by an ambuscade when proceeding to fetch water. No fresh provisions of any sort were available, as the country round the base of the hill was very poor, and what few villages existed were all deserted. The strength of the detachment was thirty Indian ranks from the Pakhangyi detachment under Lieutenant Wilson.

In March 1887 Lieutenant Lincoln and fifty Indian ranks went from Pakhangyi to occupy Lingadaw, and were reinforced shortly afterwards by forty more rifles from head-quarters. The object of this post was to prevent the incursion of dacoits from the North into the Pakhangyi district. A few days after the post was taken up, a determined night attack was made on it, and but for the prompt action of Naik Ahibaram Singh, commanding the guard, a disaster might have ensued. As it was the enemy was beaten off with heavy loss. The conduct of the Naik on this occasion was highly commended by Brigadier-General R. Low.

In March 1888 a small party from this post under Lieutenant Delamain, after a night march through the jungle, succeeded in surprising and killing a well-known leader, Boh Thaee, for whom a reward of Rs. 500 was paid by Government.

On April 3rd, 1887, a detachment, consisting of one British officer—Major E. C. Kellie—and 105 native ranks, was sent to Alon up the Chindwin. Here it remained only one week and was then marched to Mingin. The dense woods and marshy country rendered patrolling very difficult. At this time there was a scare of a Chin incursion, so a small party of the 23rd Bombay Light Infantry, mounted on transport ponies, was sent over the hills to the Toungdwenga Valley. Another party proceeded by river to Chounga and thence to the same valley, where four small posts were occupied and a regular and constant system of patrols established. Although the Chins did raid in a neighbouring valley, they retired without attacking Toungdwenga. The posts were accordingly evacuated after an occupation lasting for six weeks. The detachment returned to Pakoku on June 7th, 1887.

In December 1886, a column under Colonel J. Gatacre, 23rd Bombay Light Infantry,[1] was despatched from Pakoku against the Shwegyo-byin Montha or Kanle Prince, whose army was defeated near Myaing with great loss. The column operated in the Pakhangyi district until the end of January 1887, when it crossed over into the Yaw country to the aid of a Bengal column, under Major Stead. Colonel Gatacre, leaving the cavalry and guns in the plains and reinforced by fifty rifles, 23rd Bombay Light Infantry, crossed the Poungdoung Mountains and occupied the fortified town of Minua on February 12th with but little opposition. The column proceeded as far north as Gungaw, and then, under orders from headquarters, fell back on to Pauk and Myaing.

The Myaing village had been burnt in November 1886 before the arrival of the Pakoku column, but the situation being central, it was used as a base from which the column worked in the Pakhangyi District, and on the column being broken up, it was selected for a post to watch the dense jungles to the North-West and North in which the followers of the Kanle Prince had taken refuge. The garrison consisted at

[1] Details: 1 Squadron Cavalry; 2 guns; quarter company Sappers and Miners; 100 Rifles, Munster Fusiliers; 150 Rifles, 23rd Light Infantry.

first of two companies 23rd Bombay Light Infantry and subsequently of one-half troop Madras Cavalry, half company Mounted Infantry, and one company 23rd Light Infantry. From March to December 1887 the garrison was constantly engaged with the dacoits under Boh Nga Kwe to the South, with Bohs Thaee and Tun-u to the East and North-East, the Kanle Prince to the North, and the Leo Thugyi and Moung Yakwut to the West. The country for the most part (except to the South and South-East) was very difficult to work in, especially in the rainy season. The hills were covered with dense jungle, without roads of any description. In addition to their duties in hunting down the dacoits, the men of the 23rd were employed in making a fortified post at Myaing, in connecting Myaing with the neighbouring posts and villages by roads cut through the jungle and in the re-making of the town of Myaing. Meanwhile all the outposts had been carrying out a regular system of patrols, and from Myaing more extended expeditions had been made, in which the enemy were constantly encountered in the hills covered with dense jungle. In one of these excursions on August 6th, at Wadin, Subedar Salvadore Gabriel was shot dead while commanding the advanced guard. In him the Regiment lost an excellent Indian officer and a valued member of the shooting team. To perpetuate his memory, all ranks of the regiment subscribed and presented a monument to the Roman Catholic Church at Karachi and gave the Presidency Rifle Association a silver challenge cup to be shot for annually and known as the Salvadore Gabriel Memorial Cup.

In December 1887 a column was formed at, and despatched from, Myaing, consisting of one troop Mounted Infantry and two companies 23rd Bombay Light Infantry. The column operated in the Kyaw Valley and neighbourhood until the end of April 1888 and succeeded in quite cleaning the dacoits out of that district. In January 1888 Moung Yakwut, the chief leader of the dacoits in Yaw, was captured, tried and hanged.

Pauk was first occupied in April 1887. This was a most important point, as it was on the borders of the Yaw country, of which it was looked on as the capital. The surrounding country was much disturbed, as the famous dacoit leader, Boh Yakwut, belonged to this place and had immense influence with the people of the district. Dacoit bands were constantly roaming about in the neighbourhood and on more than one occasion threatened Pauk itself. Operations against them were extremely difficult on account of the nature of the country. The hills without roads and covered with thick jungle, and the valleys with rivers running down them and subject to inundation, rendered marching difficult, and made it impossible to surprise the enemy. There were numerous skirmishes with the dacoits, but little was effected, as, except on one occasion, they never stood. Pauk was evacuated for a short time, as it was thought it would be impracticable to hold it during the rains. On a force being marched back to re-occupy it, the dacoits were found in a strong position on the hills overhanging Pauk itself and had to be driven from one hill to the other before the town could be occupied. During the operations in the Yaw and Kyaw Valleys Pauk was an important post, as it formed the base of operations from which supplies, etc., were drawn. A detachment of the 23rd Bombay Light Infantry remained at Pauk till April 1888, under Captain E. Lawford, Madras Lancers.

Pyinchoung, an important point between Pauk and Pakoku, was held by a detachment of the regiment from April 1887 to April 1888. The principal object of this post was to maintain communication between the two places mentioned above, but a large amount of patrolling was done notwithstanding the difficult country, which was of the same nature as that surrounding Pauk.

Tebya and Kanlah were two small posts between Pyinchoung and Pakoku held for two months in 1887 to keep the line of communications open. They were eventually handed over to the Police.

A small detachment of one Subedar and twenty rifles was established on June 8th, 1887 at Poungloung, half-way between Pakoku and Pakhangyi. This part of the country had become much disturbed owing to the depredation of a band of dacoits under Boh Nga Kwe, and communication with Pakhangyi and Lingadaw had become difficult. By constant patrols in the neighbourhood confidence was soon restored. The detachment was withdrawn in November 1887. This post, like Pakhangyi, was a most unhealthy one, as it was on the border of the country flooded by the Chindwin.

In December 1887 a second column proceeded into the Yaw country commanded by Colonel J. Gatacre.[1] The force marched from Pauk on December 3rd, via Chounga, over the Poungloung Hills to Yedu, Thilni, and Gungaw. It was found that complete confidence had been restored in the valley, and the people were most willing to give all assistance in the matter of procuring transport and supplies and erecting accommodation for the troops. Although no fighting was done, the occupation of the Yaw Valley at this time was important, as it prevented the incursion of the dacoits against whom Captain Scallon was acting in the Kyaw Valley. Posts were established on the line of communications at Yedu and Thilni and garrisoned by detachments of the 23rd Bombay Light Infantry. To them was assigned the duty of patrolling the valley and passing on stores, etc., to Gungaw. The head-quarters of the column halted at Gungaw for nearly three months, and a large amount of patrolling and surveying was done. Communication was established with Captain Scallon's force in the Kyaw Valley, and at Kan with a column which had worked along the Myithu River from the Chindwin. On the return march a column, mainly composed of the 23rd Bombay Light Infantry, under Lieutenant Wilson, was sent from Thilni to Laungshe, a district not previously visited by our troops. Most friendly relations were established with the inhabitants. The column returned to Pauk about the end of February 1888, but orders were received to re-occupy the posts at Yedu and Thilni for some weeks more with detachments of the 23rd Bombay Light Infantry.

The signalling work done by the 23rd Bombay Light Infantry in Burma was extensive. As the Regiment proceeded up the Irrawady, four signallers were landed at Mimbu, where they did much useful work for four months in keeping up communication with the telegraph station on the other side of the river. In December 1886, the Shimadoung Hill in the

[1] Details: 2 guns; 50 Mounted Infantry; 1 company Rifle Brigade, and 330 Rifles, 23rd Bombay Native Light Infantry; Lieutenant Wilson as Staff Officer.

Pakhangyi District was taken up as a signalling station. The intention at first was only to connect the Pakoku-Kanle column with the head-quarters of the 4th Brigade at Myingyan, but communication was immediately established from Shimadoung with Yetagyo, Pakhangyi, Pakoku, Myaing, Toungtha, and Pagan, and later on with Lingadaw and Poungloung. Through this station columns moving in the Pekhangyi and Pakoku districts and also on the left bank of the Irrawady were able to communicate with brigade head-quarters and with one another. Being a connecting station, all messages received had to be re-transmitted and very often repeated to two or three stations. For over three months there were only four signallers to conduct all the work with two heliographs and two lamps. Afterwards two more signallers were added. The number of messages transmitted was very large, and the labour much increased by the large number of stations and the want of sufficient instruments, as, owing to the view being intercepted, the instruments had to be carried to various points of the hill to enable communication to be held with the different posts. Signalling at night was rendered difficult on account of the village lights and jungle fires which were visible owing to the lofty position on Shimadoung Hill. Although the work done was heavy, it was so well performed as to draw forth the constant praise of the late Colonel Heyland, and many other officers who communicated through the Shimadoung post. The importance of the signalling work done cannot be overestimated, as all the outposts of the Pakoku District were thus able to be in immediate communication with each other and with head-quarters, thereby facilitating concerted action throughout the district, which would have been otherwise impossible. The signallers at Shimadoung were not withdrawn until the departure of the Regiment for India in April 1888. The 23rd Bombay Light Infantry furnished signallers for the following posts in the Pakoku District in addition to Shimadoung: Pakoku, Poungloung, Pakhangyi, Lingadaw, and Myaing; and at all these stations the men were on duty for double working hours.

At the end of March 1888, on the 23rd Bombay Light Infantry being directed to embark for India, the column returned to Myaing, receiving, on its arrival there, an ovation from the inhabitants of the township, who, to the number of above a thousand, had assembled with all the village headmen to bid good-bye to the troops. The town was profusely decorated with flags and branches of trees. Triumphal arches with inscriptions in Burmese were erected, and the column headed by the local priests and headmen entered amidst the playing of bands, firing of guns and general rejoicings. For two days the men were entertained to Burmese nautches, and when, on the evening of the second day, the troops marched out of Myaing, the sorrow of the people was undisguised, and throughout the march to the river (some thirty miles) the villagers everywhere turned out to bid the men farewell.

On April 27th the Regiment concentrated at Pakoku, where it embarked on the Irrawady steamer *Sladen*, reaching Rangoon on May 5th. Here it transhipped to the *Canning* and finally disembarked at Bombay on May 17th, after nearly a year and ten months of Field Service. On the eve of their departure for India, Sir Robert Low, K.C.B., commanding the 4th Brigade, Upper Burma Field Force, issued the following appreciative Order, which speaks for itself for the fine services rendered by the 23rd in Burma :—

'The 23rd Bombay Light Infantry being about to leave the Command, the Brigadier-General wishes to put on record the services of the regiment while in Upper Burma. The regiment arrived in the summer of 1886 and has taken a prominent part in the pacification of the country. During the last year it has formed the garrison of Pakoku and the various posts attached to that Command, and it is largely owing to the ability and judgment of the Officer Commanding the Regiment, Colonel Gatacre, C.B., who has also been in command of the Pakoku District, and to the indefatigable manner in which the officers, non-commissioned officers and men of the regiment have devoted themselves to their duty that this part of the country has been reduced to comparative quiet and freedom from dacoits. The regiment has lost one native officer and nine men killed in action. During the two years it has sent only forty-eight men invalided to India, this being mainly due to careful recruiting in the past and to care taken that only efficient soldiers remained in the ranks.

'The conduct and discipline of the men have been excellent, and the turn-out of the men has been remarkable for its neatness, and the men at all times have shown a soldier-like pride in the smartness of their appearance. The Brigadier-General has noted this especially in small outposts and in parties on the line of march. Colonel Gatacre, C.B., has commanded the regiment throughout its time in Upper Burma, and the Brigadier-General congratulates him, as also the officers, non-commissioned officers, and men of the regiment on being one of the smartest Native Infantry regiments of the time.

'The Brigadier-General wishes all ranks a hearty farewell and a prosperous journey to their Presidency.'

In his report to the Commander-in-Chief, General Low remarked that 'the Regiment has seen more service than any Native Infantry Regiment in this Command, and it has come out of the trial exceptionally well. It has done brilliant service in Burma. It is remarkably well officered, and its going will be a loss to the Command'. In endorsing these remarks, Major-General White paid the 23rd the extraordinarily high compliment of adding: 'Of the many good Bombay Regiments that I have under my command here, I look upon the 23rd as the best. On all points there is evidence that the regiment is commanded and looked after with particular care and strength, and must have been trained to a high state of efficiency for long, to have stood so admirably the ordeal through which it has passed here.' On June 5th they were presented with the Indian War Medal, with clasps for Burma 1885-7, by Major-General Solly-Flood, C.B. A second clasp, inscribed 'Burmah 1887-89', was received two years later.

Appendix

DISTINCTIONS GAINED IN THE THIRD BURMESE WAR

Mentioned in Dispatches (G.O. of Govt. of India No. 434, June 16th, 1887):

LIEUT.-COLONEL J. GATACRE. 'This Officer has earned advancement by the discipline and efficiency of the fine regiment he commands. He also rendered important service in Command of a Column which advanced on Kanle and subsequently into the Yaw Valley.' Promoted to the rank of Colonel.

LIEUTENANT W. A. M. WILSON. For his conduct during the engagement at Phayagon, September 3rd, 1886. 'A well conducted affair, and creditable to Lieutenant Wilson.'

CAPTAIN E. C. KELLIE. For his conduct during the engagement with the rebels at Natogyi, September 2nd, 1886.

LIEUTENANT S. W. LINCOLN. For operations on the Chindwin in August 1886. 'The General Officer Commanding considers the operations conducted by Lieutenant Lincoln as well-conducted.'

JEMADAR SHAIKH MAHBUB. 'For the judgment and ability he displayed which on one occasion prevented the advance guard from being led into ambush.'

Companion of the Military Order of the Bath:

COLONEL J. GATACRE. (*London Gazette*, November 27th, 1887.)

Distinguished Service Order:

CAPTAIN ROBERT IRVIN SCALLON. (*London Gazette*, August 3rd, 1888.)

Third Class of the Order of Merit. For gallant conduct on the September 3rd, 1886, at the capture of Phayagon Pagoda, Upper Burma:

SUBEDAR SALVADORE GABRIEL (killed in action at Wadin, August 6th, 1887.)

No. 177. SEPOY GOPAL KANDE.
,, 237. NAIK RAMJIRAO DALVI.
,, 331. SEPOY RAJBARAO PALANDE.
,, 657. SEPOY RAMA KALE.
,, 715. SEPOY SHIVNATH AHIR.
,, 411. SEPOY SHIVNATH SINGH.
,, 432. SEPOY AMIR KHAN.

NOTE ON SIGNALLING. The 23rd were the pioneers of signalling in the Bombay Army. Colonel Gatacre started with a damaged helio picked up on the Chaman road in 1880, and afterwards obtained a complete set of Begbie Lamps, as used in the Madras Army. These proved to be a godsend in Burma, where the signallers of the 23rd did splendid work on lines of communication, keeping in touch nine stations and a number of columns.

AUTHORITIES

Regimental Records: *History of the Burmese Campaigns*, 1885-7, by Captain W. A. M. Wilson.

THE PASSING OF THE OLD BOMBAY ARMY: THE MAHSUD BLOCKADE

Chapter VIII

(1888-1902)

THE PASSING OF THE OLD BOMBAY ARMY—THE MAHSUD BLOCKADE

Soon after its return from Burma in 1888, the Regiment was moved to Kamptee, then a new station for Bombay troops, and on December 16th, 1888, at a parade of the garrison, H.R.H. the Duke of Connaught presented the Star of the Third Class Order of Merit to Havildar Ramjirao Dalvi, Naiks Gopal Kande and Rajbarao Palande, and Sepoy Amir Khan, for their gallant conduct at the storming of the Phayagon Pagoda. In doing so, His Royal Highness said: 'I have much pleasure in awarding these stars. I have to thank the Regiment in the name of the Bombay Government for the excellent service performed in Burma—service of which we are all proud. You, men, have been selected for decoration for conspicuous gallantry, that you may be an example to your comrades. I again thank you and I am proud to have met you and the gallant corps to which you belong.'

In the same year a most important reorganization was affected in the Regiment. The Duke of Connaught had determined to form a *Corps d'elite*, or Rifle Regiment, in the Bombay Army, and for this signal honour he selected the 4th, the 23rd and the 15th; he considered this to be a particularly favourable opportunity for the change, the two latter regiments having just returned from active service. The idea of the Rifle Regiment appears to have originated in the Seven Years' War in America, where fine troops, like those of Braddock, found themselves completely at the mercy of the French and their Red Indian allies, who picked them off from behind the trees at their leisure. In the American War of Independence the British had to hire continental *Jägers* to deal with the American sharpshooters. In Europe also it was necessary to take measures to cope with the French voltigeurs and tirailleurs, and the idea was taken up in England by Coote, Manningham, Moore, Napier and others. Briefly, the scheme was to form a picked body of soldiers, specially trained to work flexibly in open order, paying special attention to skirmishing, scouting, marksmanship, and individual initiative. Their uniform was to be of a neutral colour, and their weapon the rifle. Nowadays, the difference in training between the Rifle Regiment and an ordinary Infantry Regiment has been to a large extent effaced, but they still retain their distinctive uniform and drill. After several changes, it was finally decided (G.G.O. No. 14, January 13th, 1890) that the three linked battalions should be styled as under :—

4th Regiment (1st Battalion Rifle Regiment) Bombay Infantry.

23rd Regiment (2nd Battalion Rifle Regiment) Bombay Infantry.

25th Regiment (3rd Battalion Rifle Regiment) Bombay Infantry.

This rather clumsy designation the 23rd retained until 1901, when it became the 23rd Bombay Rifles (G.G.O. 837 of 1901); two years later the title was altered to the 123rd Outram's Rifles. In 1922 it became the 4th (Outram's) Battalion of the 6th Rajputana Rifles, which designation it still retains.

The Regiment having become a Rifle Regiment, permission was obtained to lodge its old colours, side by side with those hung there in 1870, in All Saint's Church, Kirkee. On March 20th, 1889, Captain Scallon, D.S.O., with an escort of six non-commissioned officers, proceeded to Poona for the purpose, and on the 25th the colours were duly received by the Bishop of Bombay from the hands of Lieut.-Colonel Ebden, 10th Bombay Native Infantry (late of the 23rd), in the presence of a large congregation. An escort of one company and the band of the 10th accompanied the colours from Poona to Kirkee.

On November 20th, 1891, the Regiment had to say good-bye to its Commandant, Colonel J. Gatacre, C.B., who was appointed Brigadier-General, Nagpur District. Born in 1841, he was offered a cadetship in the East India Company's Army at the age of sixteen, and sailed for India just as the Indian Mutiny broke out. Still a mere boy, he joined the 23rd at Poona as an Ensign in February 1858, just in time to take part in the operations against Tantia Topi. In 1860 he volunteered for service in China, and returned to the regiment at the conclusion of the campaign. He served with the 23rd in Okhamandal, in the Second Afghan War, and in the Burmese War. In the latter he fought with great distinction (despatches twice, medal with two clasps), and was promoted to the rank of Colonel, and awarded the C.B. In December 1897 he was promoted Major-General and he retired in 1902. Two years later he received the cherished honour of Colonelcy of his beloved regiment. He was made a K.C.B. in 1907 and passed away in July 1932, at the age of ninety-one, full of years and honours. He was a splendid regimental officer, devoted to his regiment and loved by all ranks. As of his brother William, it may be said of him : ' Strenuous in action and gifted with an excellent sense of efficiency and discipline, he trod his path with an unswerving devotion to duty. His simplicity of character, his great courage and powers of endurance, his manly tenderness of heart, won him the admiration of all that knew him.' He was succeeded by Major E. C. Kellie, and Captain R. I. Scallon, D.S.O. became second in command. In the same year an important change in the arming of the regiment was effected : they were issued with Martini-Henry Mark IV rifles in place of Sniders.

In 1892 the Regiment found itself in Rajkot, where some serious disturbances had broken out, and on the Proclamation Parade on January 1st of the following year, the Political Agent, Sir Charles Ollivant, said : ' We are proud in having in Rajkot a regiment which has earned such a well-deserved name as the 23rd Bombay Rifles, though we hope that during its stay it may have no more serious duty to perform than to maintain its efficiency on the parade-ground and the gymnasium.'

In 1895 a sweeping change, destined to alter entirely the character of the Regiment, came into force, and it became a 'Class Company' Regiment. In order to understand the significance of this, we must remember that the old Bombay Army, like the Bombay Police at present, was enlisted regardless of caste. As we have already seen, the original Poona Auxiliary Force was composed of Marathas and Pardeshis or Hindustanis from the Company's territories; the former were mostly allowed to depart before the battle of Kirkee. The names on the monuments of Koregaon and the Miani show that in the early days Marathas, Mahars, Rajputs, Sikhs, Mohammadans, Christians and Beni-Israel Jews were enlisted, and made good soldiers; but with the spread of the 'Pax Britannica', races which had formerly fought well were losing their martial qualities, and no longer furnished the same class of recruits. Moreover, it was thought that the introduction of Class Companies would stimulate a spirit of healthy rivalry. To appreciate the magnitude of the change, it is interesting to study the caste-return of the 23rd on its return from Burma in 1888. It was as follows :—

Christians	16
Mussalmans	
Punjab	28
Hindustan	35
Bombay Provinces	56
Marathas	
Konkani	185
Deccani	62
Pardeshis	198
Rajputs	63
Sikhs	54
Jews	2
Total	699

Under the reorganization scheme (G.O. 311 Simla, March 22nd, 1895), the composition was to be as follows :—

2 Companies Rajputs } Western Rajputana.
2 Companies Jats }
2 Companies Jats, Eastern Rajputana and Central India.
2 Companies Punjabi Mussalmans.

Lieut.-General Sir R. I. Scallon, K.C.B., K.C.I.E., D.S.O.

[*face p.* 105

The other castes were to be gradually eliminated whenever suitable opportunities occurred. About the same time, the old ' Presidency Armies ' were replaced by four Lieutenant-General's Commands, under a single Commander-in-Chief. Thus the old Bombay Army, with all its fine traditions and glorious memories, disappeared. No doubt this was absolutely necessary in the interests of efficiency and centralization, but these changes must have caused a pang of regret to many who had served under the old régime. Another change, also conducive to greater efficiency, was the introduction of the Double Company System. From May 1st, 1900, the Battalion was organized in four Double Companies ; the cadre of British officers was fixed at a Commandant, second in Command, four Double Company Commanders, four Double Company Officers, and a Quartermaster.

In 1898, Colonel Kellie retired, and was succeeded by Lieut.-Colonel R. I. Scallon, D.S.O., with Major W. S. Delamain as Second in Command. In March 1899 the Battalion left Nasirabad for Quetta, and was stationed at Pishin, with a detachment of two companies at Sibi, a spot full of old associations for them, going back to both the Afghan Wars. The devastating energy and enthusiasm of their new Commanding Officer is reflected in the excellent report they received in the following year from Major-General Sir Robert Hart, V.C., K.C.B., Commanding the Quetta District.

In November 1900 the Battalion was warned for Field Service ; a force consisting of the 23rd Bombay Rifles, a wing of the 24th Baluchistan Regiment from Loralai, and the Zhob Levy Corps, under the command of Lieut.-Colonel Scallon, was ordered to Mir Ali Khel to participate in the little frontier campaign against the Mahsuds, known as the Waziristan Blockade. Waziristan is the mountainous tract country in the North-West Frontier Province between the Tochi and Gomal rivers, and the Waziris are divided into two tribes, the Darwesh Khel and the Mahsuds. The Mahsuds, who inhabit a maze of waterless hills and ravines, are confirmed raiders, well armed, and fine fighting men. Numerous expeditions, in which their villages were burnt and their towers blown up, had proved unavailing, and the Mahsud resisted all attempts to tame him by the methods which Sir Robert Sandeman had applied so successfully to Baluchistan. It was hoped that by continually sending light columns to harry the country and by strictly blockading the passes, depredations into British territory would be stopped, and the tribesmen would be brought to their senses. The continual pressure had the desired effect. In 1902, the Mahsuds surrendered their stolen rifles and paid their fines, and Lord Curzon withdrew the garrisons from the Tochi-Gomal valleys, replacing them by tribal militia. The 23rd were posted to the Zhob section of the blockade. They were constantly employed on detachment duty, manning posts, furnishing escorts and convoys, road-making and other similar work, for nearly three years. Not much actual fighting occurred, though on one occasion, when Lieut.-Colonel Scallon was endeavouring to arrest Ahmed Khan, the murderer of Extra-Assistant Commissioner Arbab Farid Khan of Draband, Jemadar Mardan Khan, with thirty sepoys, attacked and wiped out a gang of thirteen badmashes. For this, Jemadar Mardan Khan received the thanks of the Zhob Political Agency, and was awarded a *Khillat* at a Durbar held at Fort Sandeman, on April 7th, 1902. But the campaign gave all ranks an unrivalled opportunity of working under service conditions and acquiring a thorough practical training in the field. Sir Robert Hart, who came into intimate contact with them at the time, speaks enthusiastically of their work: the Indian officers, he says, were cheery and self-reliant, and all ranks inured to heat and cold and the rough life of the Frontier, and more fit for active service than any battalion in the Quetta district; they had a thorough knowledge of outpost work, and the roads they had constructed were the best he had ever seen made by military labour. Such praise, coming from an officer of Sir Robert's reputation and experience, was something indeed to be proud of.

One of the most important tasks undertaken by the 23rd was that of convoying the Powindas through the passes. Caravans of these wild, nomadic tribes from Central Asia, with their shaggy ponies, camels and families, are a familiar sight in the North-West Frontier at the beginning of the cold weather, when they descend from the Ghazni highlands into the Punjab, and leaving their wives and children, wander all over India peddling their merchandise, returning again in March.

> When the spring-time flushes the desert grass,
> Our Kafilas wind through the Khyber Pass:
> Lean are our camels, but fat the frails
> Light are the purses, but heavy the bales,
> As the snow-bound trade of the North comes down
> To the market-square of Peshawar Town.

At the British Frontier they have to surrender their arms, to be returned to them on their homeward journey. It was usually necessary to entrust this delicate work to the local troops, such as the Zhob Levy or the Baluchi Regiment, but so readily had the 23rd picked up the ins and outs of the Zhob country that they were specially retained for the purpose. The political Agent, Zhob Militia, writing to the Government of India, said: 'The regulations of the movements of the Powinda caravans, consisting of many different clans not always on the best of terms among themselves and including subjects of Afghanistan as well as of British India, called for exercise of the greatest care, patience and tact in order to prevent serious disputes as to precedence of passage and selection of camping sites. The fact that in only one instance have disputes ended fatally, and this in territory beyond the jurisdiction of the Agency, would appear to me to reflect the highest credit upon the care of the officers charged with the Powinda transit. I would take this opportunity of bringing to your notice the great advantage that has accrued to the Zhob Agency through the military authorities having permitted the 23rd Bombay Rifles to continue their service in Zhob, when a change of regiments would have deprived the Agency of a Corps now intimately acquainted with the tribal idiosyncrasies of the Punjab-Zhob border. The construction of a practicable road through the Kanjuri Pass by the 23rd Bombay Rifles, who have previously to this made many very useful military roads on and along the Border, has operated most favourably in preventing a block in the traffic at a locality particularly liable to raids by Mahsuds.' The valuable services rendered in this connexion by Lieut.-Colonel Scallon and the troops under him were brought to the notice of the Government of India, by the Commander-in-Chief, and a letter was received from Army Head-quarters stating that 'The Government of India agree with His Excellency that Lieut.-Colonel Scallon, and the troops and levies under his command, have done excellent work, and I am to request that the acknowledgements of Government may be conveyed to them'.

Some time after, two Maliks or tribal leaders visited the Regiment in Bombay, and thanked them for the manner in which they had three times convoyed their Kafilas, with their flocks and herds, through the danger zone, under the very noses of the starving and desperate Mahsuds, without the loss of a single lamb!

The blockade being now over, the Battalion returned to Fort Sandeman, and on April 9th, 1902, marched out of Fort Sandeman en route to Bombay via Dera Ishmael Khan, reaching their destination on April 28th.

AUTHORITIES

Regimental Records and Diaries.

Blue Book on Mahsud-Waziri Operations (1902).

ADEN, THE PERSIAN GULF AND BURMA

British Officers, 123rd Outram's Rifles, 1910

Standing—Lieutenant W. Odell, Captain W. P. M. Sargent, Captain K. B. McKenzie, Lieutenant C. F. F. Moore, Captain A. K. Norris, Lieutenant R. Tilley, Captain G. E. Hardie, Captain W. R. Daniell.

Sitting—Major B. G. B. Kidd, Major R. W. C. Blair, Captain W. Greatwood, Lieut.-Colonel W. S. Delamain, D.S.O., Major E. E. Bousfield, Captain R. V. Hunt, Captain J. G. Rae.

[*face p.* 111

CHAPTER IX

(1903-1914)

ADEN, THE PERSIAN GULF AND BURMA

IN January 1903, a number of special concessions to the Indian Army were announced in honour of the coronation of His Majesty King Edward VII as Emperor of India. The designation 'Staff Corps' was abolished, and officers in future were to be designated as officers of the Indian Army. A limited number of Indian officers were to be annually appointed as Orderly Officers in attendance on His Majesty. Various gratuities were to be given as rewards for long and meritorious service, and the honorary rank of Captain was to be bestowed on Subedar Majors retiring with the First Class of the Order of British India, and of Lieutenant on other Indian Officers retiring with the same decoration. One of the first to enjoy the latter concession was Subedar Major Sardar Bahadur Shaikh Mahbub, who retired after 32 years' service, having distinguished himself greatly in the Afghan and Burmese Campaigns. He was a grand type of Indian Officer, and was granted a Jagir bringing in Rs. 400 a year. Another concession granted by His Majesty was that distinguished officers of the Indian Army might be appointed as Honorary Colonels of Indian Regiments with which they had had 'some previous distinguished association'. In the following year (July 7th, 1904) Major-General Sir John Gatacre, C.B., was appointed the first Honorary Colonel of the 123rd Outram's Rifles.

In October 1902, the Battalion was ordered to Aden, and on March 4th, 1903, it marched to Dhala in the Hinterland where it became part of the Aden column, under Colonel H. T. Hicks, C.B., to which had been entrusted the important and delicate duty of escorting and protecting the Aden Boundary Commission. The Turkish Commission had formed an entrenched Camp at Jaleli, about two miles away, with an escort of about 400 infantry, besides cavalry and guns. On March 22nd, the Turks, under instructions from the Porte, evacuated Jaleli, which Colonel Delamain proceeded to occupy. The British Ensign was then hoisted under a salute from the troops, in the presence of Brigadier-General Maitland, the members of the Commission and a large body of Arabs.

Aden had been captured by the British in 1839, but its strategic importance was enormously enhanced by the opening of the overland route, and still more, of the Suez Canal. In 1845, the Hejaz came under Turkish rule, and in 1872 Turkey occupied the Yemen. In order to define the limits between Turkish territory and that of the independent Arab tribes in political relations with Great Britain, a joint Commission of British and Turkish Officers demarcated a boundary line running from Shaikh Said north-east to Kataba, and thence to the Great Desert. This placed the whole of Southern Arabia east of the line within the British sphere of influence. Such a proceeding was rendered all the more desirable by the fact that the Arab tribes were in a state of chronic revolt against Turkey, and it was necessary in the interests of peace as well as prestige to make a military demonstration, and where necessary to conduct punitive operations, in the territories of the neighbouring chiefs. The Arabs had little liking for the presence of any foreign force, Turkish and English, and our camps were frequently fired into by night, camels stolen, and mail-bags looted, whereupon small columns were sent out to blow up towers, and destroy villages and coffee-gardens. On August 13th a detachment under Captain Shewell was attacked in force at Awabil for about five hours, the Arabs being eventually beaten off with loss. In October there was a good deal of fighting in the vicinity of a village called Suleik, where there was a post under Captain Lloyd-Jones of the 102nd Grenadiers. Captain Burton of the 123rd went to his assistance with a party of fifty rifles, and drove a number of Arabs from the neighbouring hills at the point of the bayonet; soon after, a patrol of the 102nd Grenadiers was ambushed and almost wiped out. It was then reported that 1,600 Kotaibis had collected to attack Suleik, and a strong column, consisting of two hundred rifles of the 123rd under Lieutenant G. E. Hardie, fifty rifles from the Hants Regiment, and two guns, was despatched to its relief. The enemy was quickly located and dispersed. Bayonet charges in the most gallant manner were led by Lieutenant Hardie and Captain Lloyd-Jones ; the latter was severely wounded, his thigh being smashed, and he was carried to a place of safety by Subedar Bije Singh and Havildar Ghad Singh. After this a punitive column was sent to the Kotaibi country, which destroyed a number of towers. The battalion was engaged in constant work of a similar nature and in helping to provide escorts for the members of the Commission during their work, until its return to Bombay in October 1904. Just prior to its departure, the 123rd was stationed in the friendly state of Dthalla, where it won golden opinions from the local inhabitants by its orderly behaviour and by building a number of highly useful roads. As it marched out, the Amir of Dthalla insisted on escorting it to his borders, and there, dramatically unbuckling his sword, he begged the Regiment to accept it as a memento, saying that they might forget him, but he would never forget them. The sword, a handsome blade in a silver sheath, occupies an honoured place in the Mess to-day. As they were about to leave Aden, Colonel Wahab sent the following letter to Major Delamain, Commanding Boundary Commission Escort : ' I should be very glad if you would publish an order bidding farewell to all ranks of the Escort, and expressing my thanks to all for the cheerful way in which they have worked under conditions as arduous as those of active service, but without the excitement and the chances of distinction which active service affords.'

On July 4th all ranks were overjoyed to hear that the Order of Companion of the Bath had been bestowed upon Colonel Scallon. Thereupon Colonel Scallon published the following characteristic Battalion Order :—

' The distinction granted to the Commanding Officer is intended less as a reward to him personally than an acknowledgement of His Majesty's appreciation of the good hard work done by the officers and men of the 123rd Rifles. Colonel Scallon feels very sensible that it is to the loyal way he has been supported by all ranks since he took over command, and to the reputation the officers and men have made for the 123rd Rifles, he owes this honour, and he much regrets that he alone has been rewarded. Colonel Scallon thanks one and all for the coveted honour they have obtained for him.'

Unfortunately for the Battalion, Colonel Scallon was appointed Colonel on the Staff at Bangalore on December 11th following. He had been with them in all their campaigns since 1877, and had commanded them for the last six years, and his loss was greatly felt by all ranks. He was succeeded by Lieut.-Colonel W. S. Delamain, who, on May 23rd, 1905, was awarded the Distinguished Service Order in recognition of his services during the operations in connexion with the protection of the Aden Boundary Commission, 1903-4.

The next five years were spent in the somewhat monotonous routine work of a regiment in time of peace, though the annual reports show that under Colonel Delamain, the Battalion was fully maintaining its high reputation. All, therefore, were delighted when telegraphic orders were received on January 16th, 1910 to mobilize four companies for active service in the Persian Gulf for the suppression of gun-running, in connexion with the Royal Navy. The evil effects of gun-running had been first observed in the Tirah Campaign in 1896-7, and the difficulties we have since experienced on the North-West Frontier may be largely traced to the arming of the tribes with weapons of precision ; 40,000 rifles were imported into Afghanistan in 1909, and these fetched 150 rupees a rifle on the Frontier. The trade in ammunition was in proportion. The great distribution centre for the traffic was at Maskat on the Arabian coast. It was said that the Sultan got a rupee for every rifle which passed the customs ! Although Maskat was closely watched by British cruisers, dhows would readily undertake the risking of running the blockade and landing their precious cargoes on the Mekran coast opposite, where their arrival was eagerly awaited by the Baluch Sardars. The

Sardars in their turn gave the signal to the Powindas, who were waiting with their caravans ten miles or so inland. As soon as a favourable opportunity occurred, the Powindas loaded up their camels, and were soon away on their long trek to the frontier, where the rifles were sold at an enormous profit. The Powindas were quick to see that this was far more lucrative than their annual migration to the Indian plains, and the Mekran coast soon swarmed with *Kafilas*. It was obvious that the navy alone could not tackle the situation. What was required was a small, mobile force, with sufficient transport to carry water and rations for a couple of days, on a fast steamer which could land it unexpectedly at any point on the coast. The Naval Commander-in-Chief arranged with the Government of India for the despatch of such an expedition and R.I.M.S. *Hardinge* was to convey it to the Gulf. The force consisted of 378 rank and file of the 123rd Rifles, with eight British[1] and nine Indian officers, a section of the 31st Mounted Battery, and a section of the 18th Company, Sappers and Miners, under the command of Colonel Delamain, D.S.O.

The little expedition steamed out of Bombay Harbour on January 21st. During the voyage the men were kept fit with physical drill and by practising details of disembarkation. On the 26th, acting on instructions from the flag-ship, the *Hardinge*, steaming with lights out, anchored in Kohrlash Bay, thirteen miles from Jask. The idea was to land under cover of darkness, and march to a place called Hasar, about twenty miles inland, where our Intelligence Department had located an arms depot. The place was to be surprised at daybreak, the arms destroyed, and the force return before the Afghans recovered from their surprise. Unfortunately, the mules stampeded, and the troops did not get ashore until daylight. Two Baluchis were captured and made to act as guides; the ground was soft and intersected with nullas, while all round were mud volcanoes, some of a great height, lending the scenery a most fantastic appearance. The column encountered no opposition until fire was suddenly opened on the advanced guard under Captain Daniell, from a sangar in a range of low hills about

[1] Col. W. S. Delamain, D.S.O. (Commanding); Major H. E. C. B. Nepean, Captain W. Greatwood, Captain W. R. Daniell, Lieutenant J. G. Rae.

1,500 yards away. The mountain gun came smartly into action, and under its covering fire and that of the machine guns, Captain Daniell proceeded to carry the position in gallant style, while two companies turned the enemy's flank. In the sangar were found two dead Afghans, 760 rifles and carbines, and nearly 50,000 rounds of ammunition, which were destroyed. This well planned raid was a complete surprise to the enemy, and caused much uneasiness; it forced them to concentrate instead of remaining scattered in small parties along the coast, which greatly hindered their plans. The troops returned to Jask, covering fifty-three miles in as many hours without a single man falling out, and immediately re-embarked. The *Hardinge* then proceeded to cruise up and down the Gulf; landings were made, but without result, at Guru and Sirik, the latter being a small port at the mouth of the Gaz river. A large party of Arabs, about 1,600 strong, was located at the old fort of Sarsar, eleven miles from Guru, but it was not deemed advisable to attack them with the limited forces at our disposal. Visits were paid to Maskat and Bandar Abbas and useful reconnaissances carried out, and on April 1st, the hot weather having set in, the *Hardinge* set her course for Bombay, which was reached a week later. For gallant conduct at Hasar, Subedar Major Shivnarayan Singh was admitted to the Order of British India in the Second Class, with the title of Bahadur, and Jemadar Ghulam Mohammed was awarded the Indian Distinguished Service Medal. On February 27th of the following year, the Commander-in-Chief, Sir O'Moore Creagh, V.C., while inspecting the Ahmednagar Brigade, presented their decorations to Major Shivnarayan Singh and Jemadar Ghulam Mohammed. Addressing the regiment His Excellency said:—

'I have known the Regiment for twenty-five years; it has served under me at various times, and has always acquitted itself well. When I sent the Regiment to the Persian Gulf on active service, I knew it would maintain its reputation and on meeting the Admiral of the East India Squadron a few days ago, he mentioned the good work the Regiment had done. In presenting these medals for distinguished conduct, it is not only a mark of distinction of the officers decorated, but an appreciation by the Government of the services of all ranks who took part in the operations.'

It was known that the Powindas and Baluchis would not surrender their lucrative trade easily, and reports arrived that they had collected in large numbers, so as to be able to deal with any small landing party who might try to interfere with them. Accordingly, the following year, combined operations on a larger scale against the Mekran gun-runners were planned and a force of about 1,000 men under Rear-Admiral Sir Edmund Slade, K.C.I.E., was mobilized. The landing party was again under the command of Lieut.-Colonel Delamain, D.S.O., and consisted of the 104th Wellesley's Rifles, Lieutenant W. Odell with five Indian officers and 150 non-commissioned officers and men of the 123rd, the 32nd Mountain Battery, and the 19th Company Sappers and Miners, with two sections of Field Ambulance. On April 6th the Commander-in-Chief hoisted his flag on the *Highflyer*, and the troops embarked on the R.I.M. steamers, *Hardinge* and *Northbroke*. Four days later, they arrived at the mouth of the Rapoh River on the Mekran coast, and Colonel Delamain was ordered to march to the town of Bint, seventy miles inland, which, according to the Intelligence Department, was an important centre for the arms traffic. Owing to the intense heat, the men had to march by night, and Bint was only reached at dawn on the 17th. Unfortunately, the birds had flown, and the local Khan readily made his submission. Though the troops were disappointed of a fight, there is no doubt that the expedition greatly impressed the Baluch Sardars, who thought that their inland towns at least were quite immune from attack, and must have been considerably surprised to see a large body of men suddenly descending upon them in this manner. On their return march the troops covered seventy-three miles in forty-nine hours, and were all aboard on the 20th. The expedition then sailed for Jask, where news was received of the enormities being committed by Mir Barkat Khan, the Governor of Bibujan, who had given trouble to the British before, and had, incidentally, been indulging in an orgy of gun-running. A landing was accordingly effected at Sirik, which had been visited in the previous year. It was now ascertained that Mir Barkat Khan had taken up a fortified position in a narrow gorge known as the Pashak Pass. On April 28th, Colonel Delamain moved out to attack him. Contact was made with the enemy about 9 a.m. and the troops advanced under covering fire. After a smart little action, the pass was captured. Line after line of sangars were carried with trifling loss, and by noon the enemy could be seen in full flight across the plain. This was the conclusion of operations for the season, and on May 3rd the battalion returned to Bombay, its final destination being Ahmednagar, where in 1911 it obtained the following complimentary report from Major-General Kelly, G.O.C. Ahmednagar Brigade :—

'The regiment retains its reputation as being one of the best Infantry units in the Indian Army. It is extremely well commanded. The British officers are keen and hard-working: the Native officers work satisfactorily. The spirit of *esprit de corps* runs very high. The interior economy and system of training would be hard to improve on. The men are a fine-looking lot. The wing which was engaged in stopping gun-running up the Persian Gulf earned, as I was always sure it would, the highest praise from the Admiral-in-Command.'

In October of the same year the Battalion was transferred to Rangoon, and became part of the Mandalay Brigade. In January 1912, Colonel W. S. Delamain, D.S.O., was appointed A.A.G. at Army Head-quarters, and the G.O.C. Mandalay Brigade remarked that 'the vacating Commanding Officer is an exceptionally able and devoted officer, under whom the traditional high standard of efficiency of the regiment has been well maintained'. Lieut.-Colonel H. E. C. B. Nepean succeeded him, with Major E. E. Bousfield as Second in Command. The Battalion had an uneventful period in Burma; at the Mandalay Assault-at-Arms in November 1913, it swept the board, and in December it carried out a remarkable route-march from Bhamo to Manipur, leaving Bhamo on

December 16th and arriving at Manipur on the 23rd. It was still there when, on August 4th, 1914, arrived the momentous news of the outbreak of the Great War.

AUTHORITIES

Regimental Records and Diaries. *Gun-running and the North-West Frontier*, by the Hon. Arthur Keppel (1911).

Baluchistan Agency and Secret Border Reports, 1908-10.

THE GREAT WAR: GAZA AND JUNCTION STATION

Chapter X

THE GREAT WAR: GAZA AND JUNCTION STATION

When the momentous news of the outbreak of war was received on August 4th, 1914, the battalion was at Manipur. All leave was immediately stopped, and furlough men recalled. Captain R. V. Hunt, Captain G. E. Hardie, Captain W. Odell, and Lieutenant C. F. F. Moore were sent to join the 125th Rifles, proceeding on Active Service to France; Major E. E. Bousfield, who was on leave in England preparatory to retirement, joined the same battalion at the Front. During the earlier part of the war the 123rd had not the good luck to go abroad, but it did invaluable service in training men and supplying them to the linked and other battalions. Five Indian officers and three hundred men were sent to the 125th, three Indian officers and 127 men to the 104th, and large drafts to the 58th Rifles, the 24th Punjabis, the 6th Jats, and other battalions. The depot at Mhow, under Major R. W. C. Blair, assisted by Captain J. G. Rae, passed out 2,500 recruits in fourteen months, and about eight hundred men were trained and equipped for active service. A semi-permanent camp at Happy Valley, Shillong, was set up for the purpose of giving intensive training to young soldiers. This excellent work received special recognition from the Commander-in-Chief. On June 21st, 1915, the Adjutant-General wrote: 'I am directed to request that you will convey to the 123rd Outram's Rifles, His Excellency the Commander-in-Chief's great appreciation of the very fine manner in which they have responded to the call made on them for drafts to their linked battalions and to units other than their own links. With a wide conception of the true situation which necessarily entails sacrifices by individual units for the good of the whole Army, this battalion has readily given of its best, notwithstanding that by doing so it was possibly lowering for a period its own efficiency.'

In spite of being denuded of its best officers and men, the Battalion continued to maintain its high standard. In May 1915 Brigadier-General Money reported that : 'The state of efficiency to which the young soldiers have been brought in a limited period, and under circumstances unfavourable for training, reflects great credit on Major Blair, the British and Indian officers and the instructional staff.' The officers and men thus transferred fully maintained the traditions of Outram's Rifles in the field. Lieutenant (afterwards Captain) Moore was present with the 1/4th Gurkhas at Neuve Chapelle and the second battle of Ypres, where he was severely wounded. At Neuve Chapelle, when the front of the Lahore Division was held up, he stayed out in front of the front trench under heavy fire, in order to help a wounded Gurkha and put his leg into a comfortable position ; he subsequently carried back to shelter Captain C. T. M. Hogg, who was severely wounded, and then brought in a wounded ' Tommy '. For these acts of gallantry he was decorated with the Military Cross by H.M. The King in August 1915. He rejoined the Battalion as Adjutant[1] in November 1916. Major (afterwards Lieut.-Colonel) Bousfield, who had been made Town Major at Lillers, was transferred at his own request to the 1/1st Gurkhas and was mortally wounded at Neuve Chapelle. Captain K. B. McKenzie was reported missing after Loos. Captain Odell, having served with the 125th and other regiments at Festhubert, Givenchy, Neuve Chapelle, and Loos, for which he was awarded the Military Cross, was killed in action at Sannaiyat in Mesopotamia on February 22nd, 1917.[2] Captain W. Greatwood was killed while serving with the 1st Gurkhas in an attack on the Dujaila Redoubt, March 9th, 1916. Lieutenant C. N. Harris, attached to the 125th, fell in the attack on Sannaiyat on April 22nd, 1916, when Napier's Rifles lost sixty-one per cent of their strength.[3] Captain A. K. Norris, Subedar Bije Singh, Jemadar Sabal Singh, and two drafts who joined the 104th in Mesopotamia in May 1915, were with General Townshend's force during the whole of its advance upon Bagdad. All three

[1] He died at Kantara of influenza, February 18th, 1919.
[2] See *Napier's Rifles*, p. 159.
[3] See *Napier's Rifles*, pp. 153-5.

were severely wounded at the battle of Ctesiphon. The remnant of the two drafts were in the retreat to Kut-el-Amara and the siege; and the survivors became prisoners of war on the surrender of the town in April 29th, 1916.

On March 6th, 1915 the Battalion left Manipur for Shillong, amid general regret. H.H. the Raja of Manipur wrote to the Political Agent expressing his deep sorrow that their stay had been cut short, and remarking on the exemplary conduct of the sepoys and their cordial relations with his subjects. The Political Agent, in forwarding the letter, commented upon the marked improvement effected by them in the general condition and appearance of the cantonments, the excellent discipline of the men and the smartness of their turn-out. Their stay at Shillong was brief and uneventful, and on January 2nd, 1916, they found themselves at Multan. Here news was received that a second battalion was to be formed, and stationed at Baroda with the following personnel:—

Commandant	..	Major R. W. C. Blair.
Wing Commander	..	Major H. M. P. Lord.
Adjutant ..	..	2nd Lieutenant H. Twynman, I.A.R.O.
Quartermaster	..	2nd Lieutenant H. Haslehurst, I.A.R.O.

On the occasion of their departure, a farewell function was held in the Mess, at which all officers, British and Indian, were present. The garden was illuminated by torchlight, and old and new guards formed from the respective battalions.

On August 16th, 1915 the Regiment heard with great regret that their former Commandant, Lieut.-General Sir R. I. Scallon, K.C.B., K.C.I.E., D.S.O., had been compelled to retire owing to ill-health. Born in the Mutiny year, he had joined the 23rd in 1877, and commanded them from 1898 to 1904. He had served in the Afghan War of 1879-81, the Burma Campaign, the Terah, the Mahsud Blockade and the Aden Hinterland, and was commanding the Northern Army at the time of his retirement, after thirty-eight years of strenuous soldiering. On this occasion the following regimental order was published:—

'The Commanding Officer regrets that Lieut.-General Sir R. I. Scallon, K.C.B., K.C.I.E., D.S.O., who was recently obliged to go to England on account of ill-health, has been invalided from the Service. General Scallon has asked the C.O. to express to the Regiment his sorrow at not being able to say good-bye in person. The good wishes of the whole regiment will go with its old C.O. into the retirement which he has earned so well.'

At last, on November 13th, 1916 came the long-awaited order to mobilize for Field Service in Egypt. On January 1st, 1917 the Battalion reported mobilization complete, and on the 16th it left Multan in two special trains for Karachi, embarking two days later on the *Aronda*. In command was Lieut.-Colonel G. R. Cassels, an officer who had seen much service in the Soudan and on the North-West Frontier, and had lately been G.S.O.I. at Army Head-quarters; he had been transferred from the 35th Sikhs. He succeeded Lieut.-Colonel H. E. C. B. Nepean, who had proceeded on Active Service in Mesopotamia,[1] as officiating Commandant on December 13th, 1916, and became permanent Commandant on January 13th of the following year. The other officers who accompanied the Battalion were :—

Major B. G. B. Kidd, 2nd in Command.
Major W. R. Daniell, Company Commander.
Major R. V. Hunt, Company Commander.
Major A. K. Norris, Company Commander.
Captain J. G. Rae, Company Commander.
Captain C. F. F. Moore, Adjutant.
Captain R. P. T. Ffrench, Company Officer.
Lieutenant L. A. Stuart, Quartermaster.
Lieutenant W. Aird-Smith, Attached.
Lieutenant W. A. L. James, Attached.
Lieutenant C. K. Rhodes, Attached.
Major C. A. Gill, I.M.S., Medical Officer.[2]

In addition, there were sixteen Indian Officers and 830 Indian other ranks. The Battalion landed at Suez on

[1] Lieut.-Colonel (afterwards Brigadier-General) H. E. C. B. Nepean had a distinguished career in Mesopotamia. He was five times mentioned in despatches, and was awarded the C.S.I. and C.M.G., and the C.B. in 1921. He retired in 1922.

[2] He was succeeded by Lieutenant B. H. Kamakaka, I.M.S., on June 28th. Lieutenant Kamakaka was with the Battalion throughout the War and received the M.C. for his services in the field.

Brigadier-General G. R. Cassels, C.B., D.S.O.

[face p. 126

January 18th and after spending about a fortnight in a training camp proceeded to take over the posts on the southern section of the Canal from Ismailia to Kubri. This was a quiet sector, and everyone was relieved when on July 1st they were moved up to Al Arish, in order to undergo intensive training, preparatory to going into the line. After six weeks' training, the Battalion was sent to Deri el Bela, where it was incorporated in the 234th Brigade of the 75th Division. The other battalions in the Brigade were the 1/4th D.C.L.I., the 1/4th Dorsets, and the 58th Vaughan's Rifles. After officers and N.C.O's had been into the front line trenches for instructional purposes, the Battalion in brigade moved into the line about three miles east of Gaza on September 13th. On October 1st they were in the Abbas Salient, a lively section of the front, about a mile from Gaza, with frequent shelling and raids. On October 31st the great bombardment of Gaza began. Night and day 300 guns of all calibre shelled the Turkish defences, and soon the warships on the coast joined in. Meanwhile, Lord Allenby was preparing his surprise attack upon Beersheeba, which completely upset the enemy's calculations. On the night of November 1st, the important position known as Umbrella Hill was stormed, and on the next day our troops reached Shaikh Hassan, which lies on the coast to the north-west of the town. Its capture turned the flank of the Turkish defences. The operations against Gaza culminated in the fall of Ali Muntar, the knoll dominating the enemy's line on the opposite flank, on the night of the 6th. After capturing Ali Muntar, the 75th division advanced over Fryer's Hill to Australia Hill, until they held the whole ridge running north and south on the east of the town. Next day, they took further lines of trenches on their right. Gaza was now almost encircled, and on the night of the 7th the Turks evacuated it, and by morning were in full retreat all along the line. The famous stronghold, with its intricate lines of trenches which had twice defied all attempts to take it, and was indeed, fondly supposed to be impregnable, had fallen. The town itself was in a pitiful condition. The inhabitants had been evacuated, and every scrap of woodwork torn from the houses to revet the Turkish trenches. The concentrated bombardment, together with the explosion of ammunition-dumps, had reduced what was left to a heap of ruins.

On the 9th the Battalion, after a brief period for rest and refitting, started, as part of the 75th Division, in pursuit of the flying enemy. The division marched along the main road from Gaza which joins the Jerusalem-Jaffa road at Ainwas. On its left was the 52nd Division, and protecting its right flank, an Australian mounted Division. From November 9th to the 13th the pursuit went on relentlessly, the troops covering fifteen to twenty miles a day with only brief halts, under the most trying conditions. The *Khamsin* or Sirocco was blowing, filling everyone's eyes with dust and parching their throats and men and animals suffered severely from the scarcity of water. At last, however, the weary columns emerged from the desert-belt, and begun to debouch upon the Plain of the Philistines, a broad stretch of open country between the Judean hills and the sea, with rolling pasture-lands and groves of trees. On this historic plain the armies of the Near East have marched and fought from times immemorial. Resistance from the enemy's rearguards now began to stiffen all along the line. Heavy shelling and long-range machine-gun fire caused constant delays. The villages, walled and flat-topped, and surrounded with cactus-hedges, made admirable defensive positions, especially as many of them were situated on low hills in commanding positions. Two of these, Katrah and El Mughar, were only taken by the 52nd Division after sharp fighting, and the 75th had some difficulty in clearing the Mesmiye ridge. After this, their main objective, Junction Station, was in sight. Junction Station, where the main line from Gaza meets that from Jaffa, was a vital link in the enemy's communications, and it was expected that the Turks would make a determined stand, at any rate until they had removed their rolling-stock and stores. To capture it involved a night march of nine miles through unreconnoitred country from which the enemy had not been cleared, a most difficult and hazardous operation.

Late on the evening of the 13th, after an exhausting day, the 234th Brigade had reached the village of El Mesmiye which was then held by the 3/3rd Gurkha Rifles (233rd Brigade).

The El Mesmiye ridge had only been cleared with difficulty and the 58th Vaughan's Rifles F.F. (234th Brigade) had suffered very heavy casualties. After dark, Brigadier-General Anley, Commanding the 234th Brigade, assembled the British officers of his Brigade and told them that he had received urgent orders to ' make for ' Junction Station at once. There was no information as to the enemy's position covering the station, nor were there air photos of the station or of the country near it.

The 123rd were detailed as advance guard, and General Anley explained to Lieut.-Colonel Cassels that he wished him to advance to within a mile or so of the objective, and then to send a detachment accompanied by a demolition party under a Royal Engineer officer to be detailed by Brigade Head-quarters to blow up the line to the north of the station at a point indicated on the map, with the object of preventing the Turks from removing their rolling stock. The G.O.C. did not disclose his further intention, but it is thought that this must have been to capture the station at dawn, as a night attack, without previous reconnaissance, on a position regarding which there was no information, presented obvious difficulties. While General Anley was addressing the officers, opportunity had been taken for water bottles to be filled, but there had been no time for the men to have any food. The G.O.C. said that the battalion must move off without a moment's delay.

Lieut.-Colonel Cassels asked that the demolition party might be handed over to him before he started, but he was told that the party would be sent up to him on the march. There was no time to explain the situation to the Indian officers, and at 10.30 p.m. the force moved off along the main road from Gaza to Jerusalem, the men weary and hungry, but elated at the prospect of a fight. ' A ' Company under Major Kidd had been detailed as vanguard, Lieut.-Colonel Cassels, Captain Stuart (Adjutant), Lieutenant James (Signalling Officer) and Lieutenant Ambrose (Intelligence Officer) accompanying it. There was no moon.

No opposition was encountered, and at about 2 a.m. the vanguard reached a point on a ridge about 1½ miles from the station. The station lights could be seen on the plain below and from the way they were moving about, it was evident that much activity was in progress. Lieut.-Colonel Cassels halted and sent back a message to Brigade Head-quarters giving his position and asking urgently that the demolition party, which had not yet reached him, might be sent up to him at once, as he was anxious to send off a detachment to demolish the line as ordered. After considerable delay a Staff Officer reached him who said that the G.O.C. had sent him to make quite sure of his position.

Lieut.-Colonel Cassels again asked for the demolition party and went back to Battalion Head-quarters at the head of the main guard. The road at this point was sunk between high banks. Suddenly a noise was heard on the right flank which was reminiscent of India—the tinkle of what sounded like a bullock-bell. A moment later, without further warning, a convoy of baggage animals with its escort walked over the bank right on top of the Main Guard—camels, horses, ponies, mules, donkeys! The escort, about thirty-five strong, was promptly captured, but some of them as well as some of the animals were shot, the reason for this being that fire was suddenly opened on the Battalion at short range out of the darkness and was returned.

Lieut.-Colonel Cassels then sent Major Hunt with his company to reinforce Major Kidd, as he expected that the main body of the escort would soon put in an appearance. The expectation was promptly realized. Loud cries of 'Allah! Allah!' were heard on the right flank, and it was evident that a large body of Turks were charging on to the road. Major Daniell, commanding 'D' Company asked leave to make a counter-charge. Permission being given, 'D' Company made a most gallant charge and effectually stopped the rush. Lieut.-Colonel Cassels then pushed out his remaining company under Lieutenant Huntley on Major Daniell's right flank, and the attack was beaten off. Major Kidd's and Major Hunt's companies were both attacked at the same time and were equally successful in beating off the enemy.

It was now 3 a.m., and some time later—there is unfortunately no record of the exact hour—the long-awaited demolition officer and his party arrived. He explained that the pack animal carrying the explosives had bolted, and that its recapture in the darkness had caused great delay. He was at once sent on to Major Kidd, who was ordered to proceed to effect the demolition. Shortly after Major Kidd had started, Lieut.-Colonel Cassels received an order from the G.O.C. to 'dig in', in his present position. Major Kidd's party had to make a détour of about two miles over very rough country. They pushed their way through trenches and past sentries, taking seven prisoners, but by the time they neared the railway line dawn was breaking and the enemy were on the alert. Heavy fire was opened on them and they were forced to withdraw without accomplishing their object.

Just after daybreak an aeroplane reported that the Turks were evacuating the Junction. The G.O.C. then ordered Lieut.-Colonel Cassels to send two platoons to reconnoitre. These platoons under Major Hunt entered the station without opposition except from shell fire, and captured two field guns and many prisoners. This was reported to the G.O.C., who ordered Lieut.-Colonel Cassels to take his battalion to the station and occupy it. As the Battalion advanced down the open hillside it was shelled, but no other opposition was met with. The 58th Vaughan's Rifles followed the 123rd and passing through the station, occupied some low hills to the east of it. Brigade Head-quarters and the remainder of the Brigade remained on the high ground some two miles west of the Station astride the main road. Lieut.-Colonel Cassels found it necessary to put the Station area in a state of all round defence. He had no information of the general dispositions of the enemy, and from the experience of the previous night it seemed quite probable that large bodies of the Turkish rearguards might still be in rear of him.

As a matter of fact, the Turks were in full retreat, and were vigorously pursued by two armoured cars from the 12th Light Armoured Car Battery, which inflicted heavy casualties on the enemy until turned back by shell-fire. For some unknown reason no artillery had been placed under Brigadier-General Anley's orders. This was most unfortunate, for at daybreak on the 14th two trains had been seen to leave the station and had the Artillery been up they would have had excellent targets—the trains, an enemy battery in position east of the Station, and—later—formed bodies of the enemy retreating up the Jerusalem road. It was rumoured at the time that Kress Von Kressenstein, the German Commander, was in one of the trains, but it was stated later that he had left the Station on the afternoon of the 13th by car.

The casualties of the 123rd in the night-fighting were slight, two killed and seven wounded. The Turkish losses were estimated at between fifty and sixty, fifteen of their dead being counted near the road on the morning of the 14th. The 123rd captured eight officers and about three hundred men, the most important junction in South Palestine with invaluable steam-pumping plant which the Turks had not had time to destroy, two field guns, machine shops, two engines and sixty trucks, and large stores of timber, petrol and flour.

The British officers present at the action of Junction Station were: Lieut.-Colonel Cassels, Majors Kidd, Daniell, and Hunt, Captain Stuart, Lieutenants James, Ambrose, Huntley, Aird-Smith, and Kamakaka, I.M.S.

BIBLIOGRAPHY OF THE PALESTINE CAMPAIGN

The Official History of the Great War, Egypt and Palestine, Vol. II, by Captain Cyril Falls.

A Brief Record of the Advance of the Egyptian Expeditionary Force (Cairo, 1919).

An Outline of the Egyptian and Palestine Campaigns, 1914-18, by Major-General Sir M. Bowman-Manifold (1922).

How Jerusalem was Won, by W. T. Massey (1918).

With the Turks in Palestine, by A. Aaronsohn.

The Palestine Campaigns and Their Lessons, by Colonel A. P. Wavell, C.M.G.

Regimental Records and Diaries.

Diaries and personal reminiscences supplied by Brigadier-General Cassels, C.B., D.S.O., Lieut.-Colonel B. G. B. Kidd, D.S.O., and Lieut.-Colonel R. P. T. Ffrench, M.C.

THE GREAT WAR: THE BATTLE OF NEBI SAMWIL

NEBI SAMWIL
Taken after the bombardment, November 24th, 1917
(*Courtesy of Messrs. Constable & Co.*)

[*face p.* 135

CHAPTER XI

THE BATTLE OF NEBI SAMWIL

NOVEMBER 21ST-24TH, 1917

AFTER the capture of the Junction, Lord Allenby pivoted his Army round on its right, so as to attack the famous Judean fortress which protects the City of Jerusalem. This is a great saddle-shaped block of limestone, rising to a regular crest or plateau, at a height of about 3,000 feet, with a number of spurs jutting out in a westerly direction with narrow valleys strewn with boulders. It is intersected by two tolerable roads from Jaffa to Jerusalem, running through Ramleh and Lydda (Ludd) respectively, but when they enter the hills they become mere tracks, impassable for heavy wheeled traffic. This historic fortress has been through the centuries the refuge of the Jews from their invaders, and Lord Allenby has sometimes been criticized for pressing on into this difficult terrain without halting to reorganize. But his plan was to give the Turks no rest. 'Had the attempt not been made at once', the Commander-in-Chief subsequently wrote, 'or had it been pressed home with less determination, the enemy would have had time to organize his defences in the passes, and the conquest of the plateau would have been slow, costly, and precarious.'

On November 19th the Battalion, which had been resting for three days in billets near the Jewish colony of Kuldeh, left to take part in the advance of the 75th Division, which had orders, at all costs, and as soon as possible, to get astride the Nablus road, the main line of Turkish communications, and so to cut off the enemy forces in Jerusalem from their comrades who had fled to the Plain by Sharon beyond the Nahr-El-Auja, and to compel them to surrender.

The Battalion was part of the 234th Brigade, the order of march being the 2/4th Dorsets, the 1/123rd Rifles, and the

1/4th D.C.L.I. The fourth battalion of the Brigade, the 58th Vaughan's Rifles F.F., had been attached to the 232nd Brigade. The three remaining battalions were all very weak: the British troops had been decimated by sickness, and on December 6th the Punjabi Mussalmans of the 123rd had been withdrawn under Major Kidd to mount guard over the Holy Places on the fall of Jerusalem.[1] Allowing for men with the transport, etc., each battalion had not much more than 240 rifles to take into action.

The Brigade followed the main road from Latrun to Saris, which runs through a defile with a steady rise all the way to Jerusalem. It halted for the night at Latrun under the most depressing conditions. The cold in the up-lands was intense and rain was falling in torrents. The troops were in their light summer clothing, without great coats or bedding, and had to bivouac on the open roadside. Fog and early darkness made it impossible to cover more than ten miles a day. Rations were short, as the road, a mere track, was now a quagmire in which camels stumbled and slipped, until it was completely blocked. Guns and waggons stuck fast in the mud, and had to be man-handled with desperate exertions. But the next day, wet and weary, and half-starved, the men pushed on heroically, though many of them had their boots so torn that they had to bind them up with their puttees. The opposition from the Turkish rearguards was increasing and constant halts had to be made while snipers were cleared away; fortunately the Indian troops, trained for the North-West Frontier, were adepts at this kind of warfare.

At Saris the Turks put up a determined stand on the 20th, but this place was carried at nightfall by the 232nd and 233rd Brigades, the cheering of the troops and the regimental bugle calls announcing the glad news to General Bulfin through the growing darkness.

On the 21st General Bulfin ordered the 75th to advance upon Bire, and this meant quitting the main road and following an unmetalled track to Biddu, which was, of course, impassable

[1] They rejoined the battalion on March 9th, 1918. They did excellent work in making the road from Enab to Biddu passable for heavy traffic and guns during the advance on Jerusalem, for which they received the thanks of Brigadier-General Boyce, C.R.A., and Colonel Leny, A.Q.M.G.

for artillery. The order of march was the 234th Brigade as advance guard, and the 233rd Brigade as main body; the 232nd Brigade was to make a feint-attack along the main road to Jerusalem in order to draw off the enemy.

The two brigades pushed on under heavy fire which necessitated frequent delays. Biddu was occupied without much difficulty at about 12.30 p.m., and a position was taken up awaiting further orders.

Late in the afternoon the Brigadier (Brigadier-General C. A. H. Maclean, D.S.O., who had succeeded Brigadier-General F. J. Anley, C.B., C.M.G., in command a few days earlier) came to the Battalion Head-quarters of the 123rd and told Lieut.-Colonel Cassels that he wished him to take the Nebi Samwil position with his own battalion and the 1/4th D.C.L.I. The 2/4th Dorsets were not available as they were holding an important position at the south end of the ridge connecting Biddu with Nebi Samwil. The Brigadier also said that he would place under Lieut.-Colonel Cassels' command the 3/3rd Gurkha Rifles, the 2/4th Hants from the 233rd Brigade, and the 231st Machine Gun Company. Some assistance was rendered by a Mountain Battery with camel transport, but this battery was not under Lieut.-Colonel Cassels' orders, and whenever it opened fire several enemy 5.9 and 4.2 batteries at once concentrated on it; owing to this and to the fact that the attack took place under cover of darkness, the support given was slight.

The task which confronted Lieut.-Colonel Cassels was indeed a formidable one. The hill on which the mosque of Nebi Samwil stands is 2,935 feet above sea level, and 800 feet above the surrounding country, which it commands on all sides. On it is the tomb of the Prophet Samuel, venerated by Jews, Christians and Mohammadans, and the mausoleum, with its domed mosque and lofty minaret, is a landmark for miles around. It is little more than a mile east of Biddu, with which it is connected by a ridge of high ground. It is of the utmost importance for observation purposes, as well as tactically and strategically, as it commands the Nablus road and overlooks the Holy City itself. Here, it is said, Richard Cœur de Lion stood in 1191, veiling his eyes in order that he might not gaze upon the goal he could not reach. It actually formed the extreme right flank of the Jerusalem system of defences, and Lord Allenby subsequently called it the 'key' of Jerusalem; the 75th Division commemorated its capture by taking a Key as their crest. There was, therefore, every reason to suppose that the position would be strongly held and bitterly contested.

At the time Lieut.-Colonel Cassels received his orders, the 1/4th D.C.L.I. were 1,000 yards away from him in position on the old Roman road ENE of Biddu; the position of the other two battalions was not known to him, except that they must have been considerably in rear of the Head-quarters of the 234th Brigade. He thus found himself in command of an improvised Brigade, two of the battalions of which were unknown to him, with orders to attack as soon as possible, with four weak and tired battalions and without artillery support, except for the mountain battery mentioned above, a position of very great strength, up-hill over more than a mile of very rough and rocky ground. Instructing Captain Stuart, the Adjutant, to call up the O.C. 1/4th D.C.L.I., Lieut.-Colonel Cassels proceeded to make a personal reconnaissance along the ridge. The whole area was being heavily shelled at the time; it was also under machine gun and long range rifle fire.

Lieut.-Colonel Cassels ascertained that there was one line of trenches about 800 yards long, running from the south end of the mosque along the ridge and facing West, and a second long line of trenches facing SSE at right angles to the first (forming part of the Jerusalem defences). He came to the conclusion that the best line of attack was directly on the trenches facing west. He therefore moved the 123rd to join the 1/4th D.C.L.I. and sent back word to the Officers Commanding the 3/3rd Gurkhas and 2/4th Hants to come on ahead of their battalions. He then issued orders for the attack. The objective of the 1/4th D.C.L.I. was from the south corner of the mosque enclosure to a track which could be clearly seen on the hillside; that of the 123rd from the track inclusive to the south end of the trench. The 1/4th Hants and half of the 3/3rd Gurkhas were to support the attacking battalions, with the rest of the 3/3rd Gurkhas to form the reserve. The last two battalions could now be seen coming up in the distance and as Lieut.-Colonel Cassels was anxious to attack at once, in order to take advantage of such light as was left and to avoid the chance of reinforcements coming out from Jerusalem, he ordered the 1/4th D.C.L.I. and the 123rd not to wait for the supporting battalions.

At 5.15 p.m. the attacking battalions moved off with Major Daniell in command of the 123rd. Lieut.-Colonel Cassels kept Captain Stuart with him as Staff Officer. The supporting battalions on arrival were sent on after the attacking battalions. The troops came under fire at once, but thanks to the failing light this did little damage. The Battalion advanced in lines of platoons in extended order, 'C' Company on the left, 'D' Company on the right, with 'B' Company and Battalion Head-quarters in support. Owing, however, to the darkness and the nature of the ground, it became more and more difficult for the different units to keep touch; three platoons under Lieutenant Ambrose missed their way and bore off to the right.

At 1,200 yards the fire became more intense. It came from a position known later as the Old Redoubt. Some time later Jemadars Arjun Ram and Bhim Singh, of the leading platoons of 'C' and 'D' Companies, joining up with the leading platoons of the 1/4th D.C.L.I. on the left, rushed the redoubt in brilliant style, bayoneting the machine-gun team, including a German officer, and capturing the gun. The advance was then continued until both battalions had made good their objectives. This was reported to Lieut.-Colonel Cassels, a further report being sent later to the effect that the mosque and its surroundings appeared to be unoccupied. Lieut.-Colonel Cassels, who had been following up with the reserve, made a personal reconnaissance of the mosque, accompanied by Captain Stuart and Lieut.-Colonel Smith of the 1/4th D.C.L.I. He then ordered the Officer Commanding 3/3rd Gurkhas to occupy it and to take up a defensive line in the outskirts beyond. By 11.30 p.m. he was able to report to Brigade Head-quarters that the whole position was in his hands. In reality it had been secured much earlier, but owing to darkness and difficulties of communication, Lieut.-Colonel Cassels was not aware of this. In the meantime he had been rejoined by Lieutenant Ambrose and the three missing platoons, who had had a series of extraordinary adventures. Lieutenant Ambrose had passed the southern flank of the Old Redoubt in the darkness, and had penetrated a good 1,000 yards beyond the objective, going down a *wadi* and up a hill in face of heavy fire. He took three lines of trenches, killed thirty Turks and captured a machine gun and thirty prisoners. Then, finding himself outnumbered, outflanked and in danger of being cut off, he coolly withdrew his men and eventually found his way back. For his action he was awarded the Military Cross.

Our total losses in the capture of Nebi Samwil were only thirty-six, while the enemy must have suffered severely, for a formed body was caught at point-blank range by a Lewis gun of the D.C.L.I.'s and dispersed. We captured forty-two prisoners and two machine-guns. The Turks were entirely taken by surprise before they had time to bring up their night reinforcements, and the failing light was a great advantage to the attack. The mosque, surrounded by buildings with walled enclosures, had the appearance of a small village. The whole position was one of great strength; it could not have been taken in broad daylight except by a strong force with adequate artillery support. The British officers of the 123rd who took part in the attack on Nebi Samwil were Lieut.-Colonel Cassels, Major Daniell, Major Hunt, Captain Stuart, Lieutenant James, Lieutenant Ambrose, Lieutenant Kamakaka, I.M.S., and Lieutenant Aird-Smith (with transport).

Immediately he heard that Nebi Samwil was in our hands. Lieut.-Colonel Cassels proceeded to put it into a state of defence, He disposed his force as follows :—

West, north and east of the Mosque and slightly forward of it : 3/3rd Gurkhas.

Prolonging the line to the south : 1/4th D.C.L.I.

Prolonging the line SSW (in Old Redoubt) : 123rd Rifles.

In close support of the 1/4th D.C.L.I. : 2/4th Hants (less one platoon in reserve).

Early the following morning the Brigadier arrived and took over from Lieut.-Colonel Cassels, who resumed command of the 123rd.

It was not, however, to be supposed that the enemy would surrender the Key of Jerusalem without a struggle. To take Nebi Samwil was one matter; to hold it quite another. At 7.45 a.m. the Turks opened an intense bombardment on the position from a large number of guns from every point of the compass, a battery from the neighbouring elevation of El Jib taking it in enfilade. Nebi Samwil was a landmark for miles around and every Turkish battery had the range to a yard. To this our troops could not reply. Our artillery was still miles behind, the gunners struggling with eight horse teams and constant man-handling to get their guns over the boulder-strewn tracks.

The bombardment went on all day without a pause. The whole hillside was plastered with shells. All telephone wires were quickly severed. The only shelter was behind a ridge about 200 yards south of the mosque, where Brigade and Battalion Head-quarters were located. The Old Redoubt, held by the 123rd, afforded practically no cover and the troops in the mosque-enclosure and linking up with the 123rd were not much better off. The scene in the enclosure itself was almost indescribable; the air was filled with smoke and dust and flying splinters of stone, and the din of bursting shells, the rap-rap of the machine guns in the minaret, and the constant bursts of rapid fire was appalling. The mosque was simply crumbling to pieces; the roof was blown in, and huge stones fell upon the wounded inside, who had to be hastily removed to the crypt. The shrine with its massive silver lanterns was demolished by a direct hit; the minarets were repeatedly struck and came crashing down upon the occupants of the enclosure. Under the cover of this bombardment, the Turks launched a series of attacks. About noon bodies of the enemy appeared from the direction of Beit Iksa, but were broken up by the fire of the 123rd, the D.C.L.I. and Gurkhas. A second attack, this time from Beit Hanun, was pushed home with great gallantry, the enemy climbing the hillside and actually reaching the crest when they were counter-attacked with the bayonet and hurled back by two companies of the 123rd. A third attack at 2 p.m. reached the summit of the hill on three sides, and the Brigadier was forced to call up all available reserves to meet it. The enemy when driven off took up positions among the boulders on the slopes whence they maintained a harassing rifle fire upon the defenders. At 3.30 p.m. came the most violent assault of all. After the bombardment had reached almost hurricane intensity, the enemy once more came on in great strength. The mosque was almost completely surrounded. A party of Turks effected a lodgment in the Old Redoubt, a vital position in our line; but Majors Daniell and Hunt called upon their men for a supreme effort and they were swept out at the point of the bayonet. Major Daniell was severely wounded, and Major Hunt reported that he had only a handful of men left and that his ammunition was running very low. A platoon of the D.C.L.I. with fresh supplies of ammunition shortly afterwards came to his assistance.

At the same time a bitter struggle was raging at the mosque itself held by the Gurkhas. A party of Turks forced their way up to the door of the courtyard, led by a German officer, but Lieut.-Colonel Channer bayoneted the officer and managed to slam the door in their faces; then, leading the Gurkhas through a back entrance, he took the Turks unexpectedly in flank. There was a short, sharp scuffle. The Gurkhas got home with their kukris, drove the enemy down the terrace and then proceeded to exterminate them by hurling down boulders on top of them. Thus the great attack was beaten off, but it was obvious that flesh and blood could not hold out much longer. The place was a shambles. The men were reduced to a mere handful. They had been fighting all day and were completely exhausted. Ammunition was running low and all the machine guns were out of action. Still there was no question of retreat and all were preparing to sell their lives as dearly as possible, when British troops were seen to be advancing from the direction of Biddu. The 156th Brigade (52nd Division) had been ordered to go to the relief of the defenders of Nebi Samwil. Advancing by forced marches without transport they had reached Biddu at 2 p.m. Here they could plainly see that fierce fighting was in progress and at once came into action. By nightfall the Turks were driven back along the whole front of the position.

That night the 123rd were withdrawn from the line and held in reserve some 300 yards from the Head-quarters of the 156th Brigade, where they remained throughout the next day (23rd). There was very little cover and the shelling throughout the day was continuous. One enemy battery scored a direct hit on the Battalion, killing two Indian other ranks and wounding twenty-five more and five mules. Late in the afternoon Lieutenant Huntley joined the Battalion with a small reinforcement. That night the Battalion was withdrawn to the vicinity of Biddu, but early next morning (24th) orders were received that they were to return to Nebi Samwil, where reinforcements were urgently needed. The strength of the Battalion in British officers and Indian other ranks was by now reduced to seven (Lieut.-Colonel Cassels, Captain Stuart, Lieutenants James, Ambrose, Huntley, Aird-Smith and Kamakaka, I.M.S.), and ninety respectively, Major Hunt having to be evacuated with a sharp attack of fever.

On reporting at Head-quarters, 156th Brigade, Lieut.-Colonel Cassels was told that his men were to take over the defence of the mosque. He led them up the hill to the mosque and placed Lieutenant Ambrose in command of one half-company and Lieutenant Huntley in command of the other. By this time the mosque was reduced to a chaotic ruin. Throughout the day it was heavily shelled by guns of every calibre and by trench-mortars, and the noise inside baffled description. During the day, however, no enemy infantry attack developed and the casualties of the 123rd were slight, and at about 10.30 p.m. they were finally withdrawn and marched through the night to a position south of Biddu.

The total casualties of the 234th Brigade in the attack and defence of Nebi Samwil were 567, or over fifty per cent of its strength. The 3/3rd Gurkhas lost 216 men and had only one British officer and sixteen Indian other ranks unwounded. The 123rd had 209 casualties, including Major Daniell and five Indian officers. Their greatest loss, however, was the gallant Major Daniell, who died of wounds at Kantara on December 1st.

In announcing his death, Lieut.-Colonel Cassels wrote: 'The Commanding Officer feels assured that all ranks will share his great personal sorrow in the loss which the Regiment has sustained by the death of this gallant officer. Major Daniell's kindness and consideration towards all with whom he came into personal contact, his resourcefulness in every emergency, his coolness and courage in the face of danger, have endeared him to all ranks. His gallantry in the action where he sustained the wounds from which he died was conspicuous and will never be forgotten. He leaves behind him a record which every man in the Regiment will do well to emulate—a record of unflinching devotion to duty, and he will long be remembered as a fine soldier, a gallant gentleman, and the friend of all who knew him.'

Such is the story of the Battle of Nebi Samwil, one of the fiercest and most desperate actions in which the Indian Army has ever taken part and one which will be remembered for ever with pride by all ranks of Outram's Rifles. It may be mentioned that up to the November 30th the Turks made repeated and unsuccessful attempts to recapture Nebi Samwil. In his despatch Lord Allenby wrote: 'These attempts cost the Turks very dearly. We took 750 prisoners between November 27th and 30th and the enemy's losses in killed and wounded were undoubtedly very heavy. His attacks in no way affected our position or impeded the progress of our preparations.' Two days after the 123rd left Nebi Samwil, Major-General Palin, Commanding the 75th Division, addressed the Battalion and congratulated them on their performance. Later on he again signified his appreciation in writing as follows:—

'I wish to convey in writing to all ranks my appreciation of the excellent work performed by the 75th Division from the day when it marched out of the Sheikh Abbas Salient (Gaza) in pursuit of the Turk. I have already expressed my opinion verbally to G.O.C.'s Brigades and O.C.'s Battalions, but I wish it to be known to all ranks of the Division. The capture of Junction Station and Latrun, the forcing of the strong positions astride the Jerusalem road leading to Enab, the relentless pursuit of the Turk culminating in the fighting which secured and retained Nebi Samwil in the face of numerically superior artillery, are all facts of which any Division may well be proud.

'Bad weather and reduced rations, brought up with great difficulty, rendered the success achieved by the men in their summer clothing still more creditable.

'I feel sure that the name won by the Division by the dash, hard fighting, and loyal co-operation of all its units in the recent operations, will inspire all ranks in the future, as in the past, to do their utmost to add to the reputation of the 75th Division.'

P. PALIN, Major-General,
Commanding 75th Division.

9.1.18.

APPENDIXES

CASUALTIES IN THE BATTLE OF NEBI SAMWIL.

Killed .. .. Major W. R. Daniell.
Wounded .. .. Subedar Major Bhura Ram.
Jemadar Pailad Ram.
Jemadar Bolad Ram.
Jemadar Bhim Singh.
Jemadar Harnath Singh.

Indian Other Ranks: Killed, 69; Wounded, 138; Died of Wounds, 2; Total, 209, killed, wounded and missing, out of approximately 350.

AWARDS

Military Cross .. Lieutenant R. D. Ambrose.
I.O.M. .. .. .. Subedar Major Bhura Ram.
Jemadar Harnath Singh.
Jemadar Bhim Singh.
Jemadar Hans Ram.
Jemadar Arjun Ram.
I.D.S.M. .. .. Naik Ganpat Singh.
L.N. Ram Kawar Singh.
L.N. Sheoji Singh.
L.N. Sanwant Ram.
Rfm. Momray Ram.
Rfm. Sheochand Ram.
Rfm. Ganpat Singh.
Rfm. Ladu Khan.

ACTIONS FOR WHICH SOME OF THE ABOVE AWARDS WERE MADE

SUBEDAR MAJOR BHURA RAM. On November 21st-22nd, 1917 at Nebi Samwil displayed great courage under heavy shell-fire until wounded on 22nd. His example animated all ranks.

SUBEDAR HARNATH SINGH. On the night of November 21st-22nd, 1917, very gallantly controlled and led his men during the attack on the right of Nebi Samwil, and helped to capture a machine gun. He charged three separate lines of trenches, and very ably helped to withdraw his men when ordered. On November 22nd behaved gallantly in holding on to his trenches under heavy shell and M.G. fire. Since wounded.

LANCE-NAIK RAM KAWAR SINGH. On the night of November 21st-22nd, 1917, during the attack on Nebi Samwil took two other signallers and charged a party of the enemy who were firing into D. Company from the right flank. The result of this bold action was that the whole party of eleven Turks threw down their weapons and surrendered.

'THE FIRST VIEW OF THE HOLY CITY'

There is an interesting sequel to the battle of Nebi Samwil. After the Armistice, Mr. James McBey, who had been the official artist with the Egyptian Expeditionary Force, was commissioned to paint a picture entitled 'The First View Of The Holy City', for the Imperial War Museum. It was thought that this first view had been obtained from Nebi Samwil, and Mr. McBey was asked by the authorities to get into touch with Lieut.-Colonel Cassels regarding the details. This he did.

Later Lieut.-Colonel Cassels, who wished to give a copy of this picture to the Officers' Mess of the 123rd, asked his wife, who was then in England, to arrange matters with Mr. McBey. (Mrs. Cassels had visited the battlefields round Jerusalem with her husband in the Spring of 1920.) She went to see Mr. McBey and he showed her the picture which he had painted. She suggested certain modifications, which were embodied in the copy which now hangs in the Mess.

The original is to be seen in the Imperial Institute, South Kensington.

Map 10
Battle of
NABI SAMWEIL.
21st November, 1917.
Capture of Nabi Samweil by 75th Division and
Attack on Beitunye by Yeo. Mtd Div
(Yeo. Mtd Div is shown at 3 pm. 75th and 52nd Divisions at 6 p.m.)
Defences of JERUSALEM under construction are shown by green dashes
Scale of Yards
Heights in Feet
EL JIB from the south-west.
NABI SAMWEIL from the west (hill 2878)
YEO. Mtd DIV.
24TH DIV.
52ND DIV.
75TH DIV.
53RD DIV.
XXI
XX
III.
JERUSALEM
Beitunye
Beit Ur el Foqa
Beit Ur el Tahta
Beit Liqya
Beit Duqqu
Beit Anan
Biddu
El Jib
Bir Nebala
Jedire
Qalandye
Ram Allah
Bire
Beitin
Er Ram
Jaba
Hizme
Anata
Shafat
Lifta
Qaloniye
El Qastal
Qaryet el Enab
Beit Surik
Beit Iksa
Deir Yesin
Et Tire
Qatanne
Beit Nuba
Kharbetha
155 Bde
156 Bde
8 Mtd Bde
22 Mtd Bde
5 Mtd Bde
H.Q. YEO Mtd DIV.
Map compiled by Historical Section Military Branch
Ordnance Survey 1929

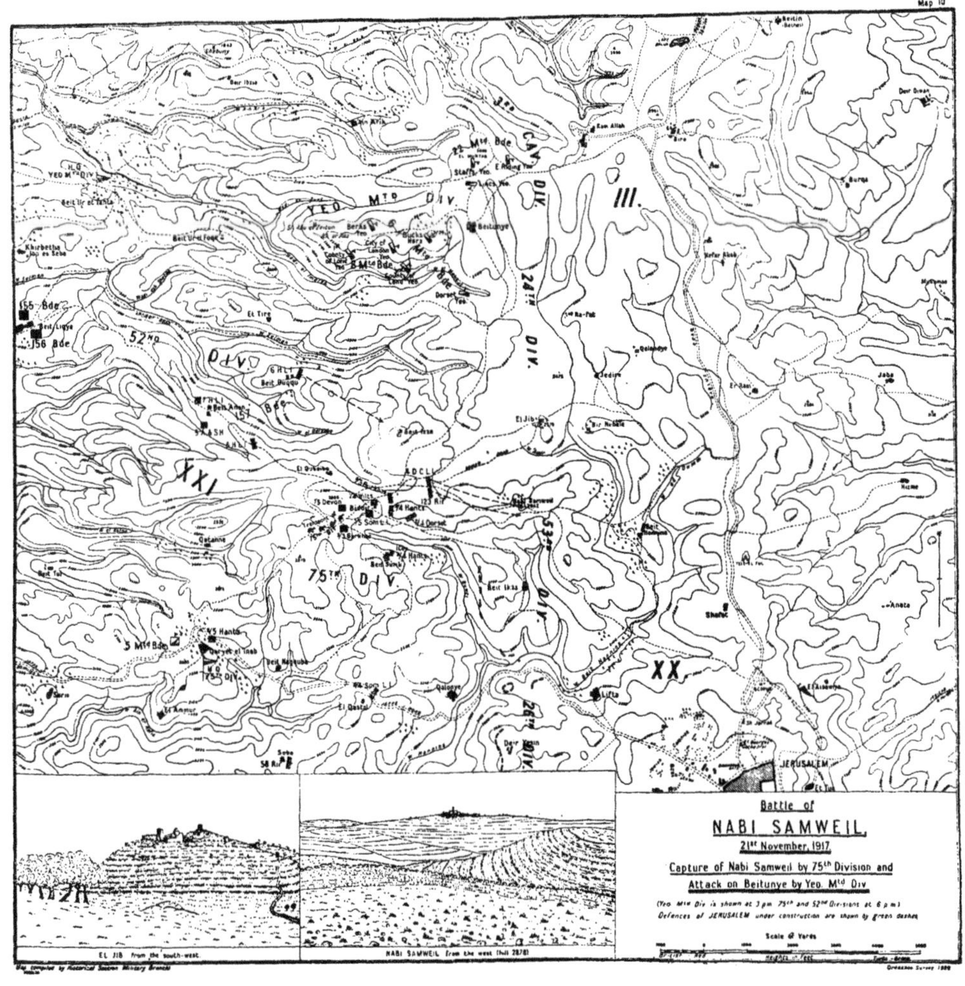

[*face p.* 146

THE GREAT WAR: THREE BUSHES HILL

Chapter XII

THE GREAT WAR: THREE BUSHES HILL

After the battle of Nebi Samwil, the Regiment went into camp at Ramleh to rest and refit. Gaps in the ranks had to be filled, and fresh drafts trained. Owing to the fact that the Punjabi Mussalmans were much over strength, and to the difficulty of obtaining Jat and Rajput drafts, the composition was temporarily altered to two companies of Punjabi Mussalmans, one company of Rajputs and one company of Rajputana Jats. During the winter the Battalion was in quiet sectors of the line at Hamid and Tireh, and was occupied in road-making, and when circumstances permitted, regimental training in preparation for the spring offensive. During the winter months, very little fighting was possible on either side. Torrential rains filled the *wadis*, converting the rivers into unfordable torrents and the low-lying lands into impassable swamps. Unmetalled roads quickly became unfit for traffic, and the railway lines were even washed away. The problem of supplying the large force now in the field soon became one of the utmost difficulty.

With the coming of the spring, however, Lord Allenby decided to take the offensive in the coastal sector. His object was to break the Turkish defences in the Ra-fat area, to such an extent as to allow the Australian Mounted Division, which was to be concealed in the orchards behind the line, to ride through the gap, following the Qualquilye—El-Tireh road, and make a surprise attack on Tulkaram. Tulkaram, the Head-quarters of the Turkish Eighth Army, was an important nerve centre. It is both a railway junction and the meeting place of the main roads running north to Haifa and east to Nablus, and its capture would paralyse their communications. The enemy were well aware of the importance of keeping this sector intact, and had stationed in the line the German ' Pasha ' division, which had recently been reinforced by two battalions, with machine-gun and artillery units.

The task was entrusted to the 21st Corps under General Bulfin, which consisted of the newly-arrived 7th Indian Division on the left, the 54th Division in the centre and the 75th Division on the right. The attack was planned to take place on a front of fifteen miles from the Mediterranean to a point east of Ra-fat on the Wadi Ballut. The 75th Division found itself opposite a ridge of high ground running from Three Bushes Hill on the left to Tin Hat Hill on the right and including the fortified villages of Ra-fat and Berukin. The division was disposed as follows:—the 234th Brigade on the left, the 233rd Brigade in the centre, and the 232nd Brigade, linking up with the 10th Division on the right. The 234th Brigade consisted of the 1/4th D.C.L.I., the 2/4th Dorsets, the 58th Vaughan's Rifles and the 123rd Outram's Rifles: their objective was Three Bushes Hill, a flat-topped ridge about 830 feet high, sloping away sharply to the north. This had to be cleared of enemy and held, so as to protect the flank of the 233rd Brigade during its advance.

On March 13th, the Regiment, in conjunction with the 1/4th D.C.L.I. and the 2/4th Dorsets, as a prelimimary operation, attacked the Turkish line above Wadi Sahury, meeting with very little opposition. The Brigade then crossed the Wadi Ballut and linked up with the 54th Division on their left. On the 18th the Regiment was taken out of the line for a rest, and was moved up again on the 27th as reserve to the Brigade centre. On April 8th Lieut.-Colonel Cassels was ordered to take command of the 234th Brigade from Brigadier General Lane, who had gone sick, and handed over the Regiment to Major Kidd. The officers with the Regiment were now as follows:—

Major B. G. B. Kidd ..	Officiating Commandant.
Major R. V. Hunt ..	O.C. 'B' Company.
Major A. K. Norris ..	O.C. 'C' Company. (Wounded April 9th.)
Captain C. F. F. Moore, M.C.	Adjutant.
Captain R. P. T. Ffrench	O.C. 'D' Company. (Wounded April 11th.)
Lieutenant W. Aird-Smith	Transport Officer. (Wounded April 11th.)
Lieutenant W. A. L. James	Signalling Officer.

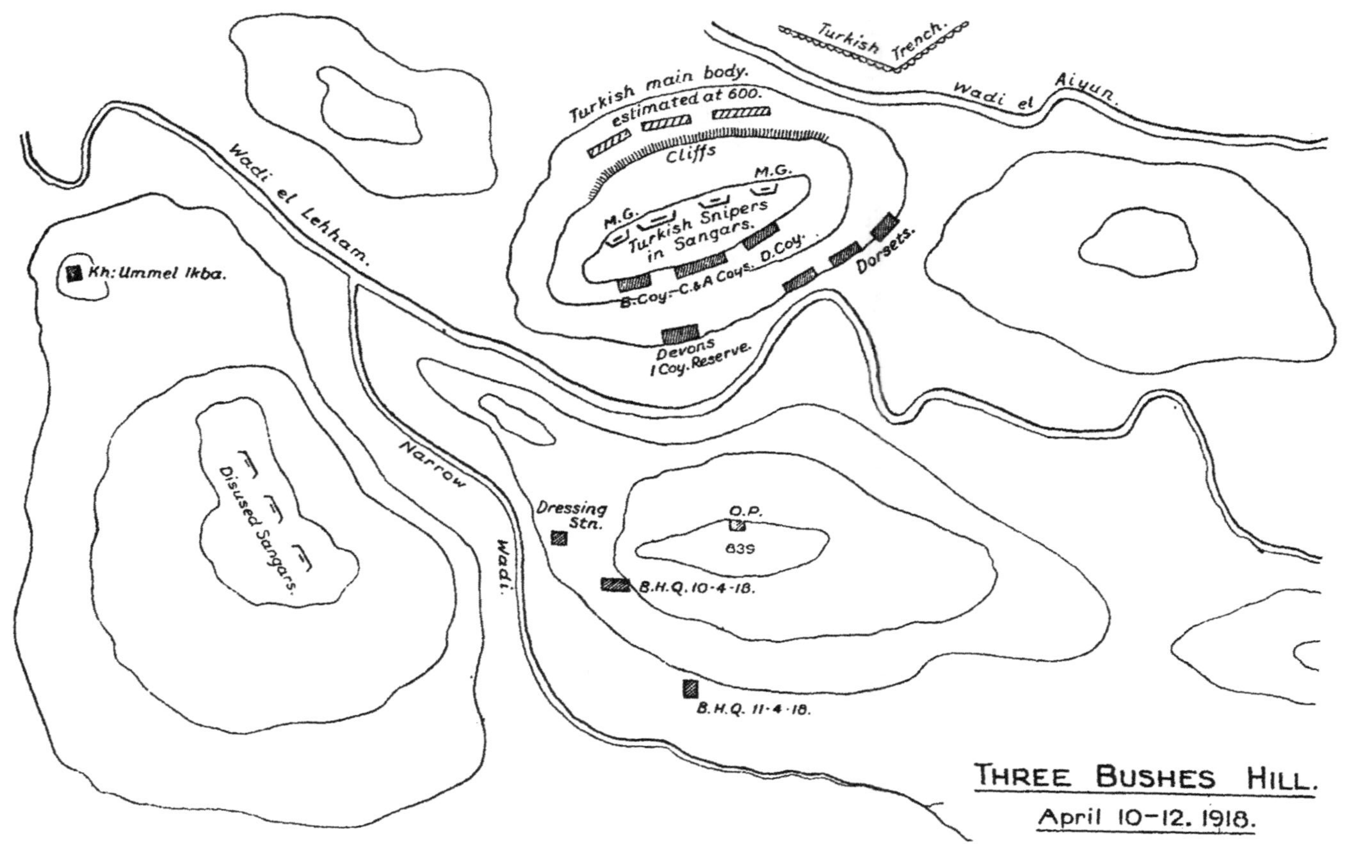

THREE BUSHES HILL.

April 10–12. 1918.

Lieutenant	Darwell ..	O.C. 'A' Company.
Lieutenant	R. D. Ambrose, M.C.	Intelligence Officer.
Lieutenant	A. Morrison ..	Quartermaster.
Lieutenant	H. S. Peppin ..	Company Officer.
Lieutenant	B. H. Kamakaka, I.M.S.	Medical Officer.

The battle started at 5.10 a.m. on April 9th, the three brigades advancing in line against their objectives. At first the Turks appeared to be taken by surprise, but the resistance stiffened as the attack progressed. On the right flank of the division, the 232nd Brigade, on emerging from the deep ravine of the Wadi Ballut came under concentrated fire from Tin Hat Hill and was unable to make any progress. The fight on this flank was very severe, and it was not until 4 o'clock that our men reached the outskirts of Berukin. Meanwhile, at 7.45 a.m. Outram's Rifles had received orders to move to the support of the 2/4th Dorsets, who were attacking Three Bushes Hill. At 10.45 a.m. Major Hunt was ordered to reinforce the firing line with 'B' Company and at once came up under heavy shell and rifle-fire, and about the same time Major Norris was wounded by a shell. At noon, 'A' Company under Lieutenant Darwell was also ordered to reinforce the Dorsets, and had two men killed while moving up. An hour later, 'B' Company was ordered to take over the left of the line, while 'A' Company was to report to the D.C.L.I. on our right flank. The Turks had established themselves just below the crest of the hill which formed a kind of No Man's Land between them. During the whole of April 10th, fierce and bitter fighting raged all along the line, both divisions being heavily engaged along their entire front. In the streets of Berukin, Englishman and Turk strove hand to hand with bomb and bayonet among the houses and in the narrow lanes. The roar of the artillery and the rattle of musketry was incessant. In the centre, the Gurkhas had hacked their way to Mogg Ridge, and Rifleman Karanbahadur Khan had won the Victoria Cross by capturing a machine-gun single-handed and turning it with devastating effect upon his opponents. But the result still hung in the balance. On the afternoon of the 10th Major Kidd received orders from Lieut.-Colonel Cassels to relieve the Dorsets after dark, and before dawn to clear the enemy off the hill, and take up such a line as would prevent him from reoccupying it. The line was to be held strongly at night, and lightly with well-concealed observation posts by day, in view of probable hostile shelling, the remainder of the troops being drawn well back under cover. The hill was flat at the top and the Dorsets now reported that the Turks held the northern part of it, the nearest point of the enemy's line being within sixty yards, and plenty of snipers within 150 yards. Major Kidd thereupon proceeded with Captain Moore, Captain Ffrench, and Lieutenant James (Signalling Officer) to the Dorsets' Head-quarters to reconnoitre the ground and make the necessary arrangements. After pointing out to Captain Moore and Captain Ffrench the line the Companies were to take up, Major Kidd sent them back to bring up the regiment at dusk, while he himself went forward to reconnoitre No Man's Land. He then went back to Battle Head-quarters and issued orders for attack at dawn. The relief was carried out under great difficulties. The enemy, only a stone's throw away, was constantly on the alert. In spite of this, however, the Dorsets were successfully withdrawn by 2 a.m. on the 11th. Two hours later the 123rd ' went over the top ' in the dim light of the dawn. ' B,' ' C ' and ' D ' Companies were in the firing line and ' A ' Company in reserve. They were met by intense machine-gun and rifle fire at point-blank range. But Outram's Rifles were not to be denied. By a brilliant charge, though officers and men were dropping fast, they swept the Turk out of the *sangars* and after a fierce struggle drove him down the slope ; as he retired, the South African Field Artillery, ranging on the flash of his rifles, put down a barrage which took a heavy toll. Captain Ffrench led his men with conspicuous gallantry ; though wounded in two places by a bomb, he refused to fall out and continued to inspire all around him by his courageous example. Subedar Harnath Singh, finding that all officers were down, took charge of two platoons, rallied them, and led them again and again against the enemy, with utter disregard of his personal safety. Jemadar Nanig Ram, seeing that Major Hunt was fully occupied in commanding the firing line and all Indian officers had become casualties, took command of ' B ' Company at a most critical moment, when the Turks in force had approached within bombing distance. He proceeded to break up the Turkish counter-attack with bombs and rifle-fire, rallying the men again and again, and keeping the line intact. Rifleman Tota Ram kept his Lewis gun in action until surrounded by the enemy, and brought it away in safety. For these gallant deeds Captain Ffrench and Subedar Harnath Singh received the M.C., and Jemadar Nanig Ram and Rifleman Tota Ram the I.O.M. and the I.D.S.M. respectively. During the attack and the subsequent heavy shelling, the Regiment suffered ninety casualties.

It was now broad daylight, and the men were withdrawn to their day positions on the southern slopes, leaving sentries in the observation posts and *sangars* on the crest. April 11th passed quietly, save for intermittent shelling and sniping, but at 6 p.m., just as the night reliefs were preparing to come up, the whole hill-top was subjected to a heavy bombardment from 5.9 inch and 4.2 inch with high explosive and shrapnel, causing many casualties to the men in the advanced positions and putting two Lewis guns out of action. Our strength in the line was reduced to 320. At 6.30 p.m., as 'B', 'C' and 'D' Companies were moving forward to take up their positions for the night, they were met by a machine-gun barrage and heavy rifle-fire and sustained severe losses in attempting to break through it. Creeping up under cover of the barrage, a large party of the enemy, estimated at 500, occupied our newly made *sangars*, which they had made untenable by the rain of shells, and our men were obliged to fall back to the original line of *sangars* on the southern crest of the hill. The enemy then launched a powerful attack, which was driven off with the assistance of a barrage. Lieutenant Aird-Smith was wounded.

The position was now critical. The enemy was established on the other side of the crest, scarcely 100 yards away. Our line was dangerously thin, the men were without food or water, and ammunition was running low. All telephone communications were cut. Major Hunt, who was commanding the firing line, managed to get a runner through to Battalion Head-quarters asking for reinforcements, as 'B' and 'D' Companies had had forty or fifty casualties, including Lieutenant Aird-Smith and five Indian officers, and 'C' Company in the centre had been reduced to fifty men. 'A' Company, the last remaining reserve, was put into the line, and presently Lieutenant Ambrose, followed later by Lieutenants James and Morrison, managed to struggle in wonderful style up the hillside to the firing line with ammunition, food and water. This was carried on camels and mules, which they had to lead over slippery rocks and up and down precipitous *wadis*, under fire which inflicted several casualties. At 8.30 p.m. three companies of the Dorsets arrived and were sent in support to the line, with extra ammunition and bombs. Two hours later, the enemy, who included a number of Germans, opened rapid fire at point-blank range and again attacked our *sangars*, using egg-bombs and heavy grenades. Major Kidd was forced to ask Captain Maasdorf, of the South African Artillery, to shorten the barrage to fifty yards beyond our line. The gunners nobly responded, and the Turks were once more forced to return to cover. April 12th, in this sector, passed comparatively quietly: the Turks had been too badly mauled to attempt any further attacks, and that night the Battalion was relieved by the 2/4th Devons. They had suffered 237 casualties between April 9th and 11th, more than half those of the whole brigade.

On coming out of the line, they received the following inspiring message:—

'Corps Commander sends his congratulations to the 123rd Outram's Rifles for their dash in clearing the enemy off Three Bushes Ridge. Divisional Commander adds his congratulations on this well-earned recognition, and feels certain that the fine work of the battalion will be appreciated.'

Meanwhile, General Bulfin had come and made a personal inspection of the battlefield. It was obvious that the original plan of rupturing the enemy's line and penetrating as far as Tulkaram was now out of the question. In these rough hills, machine-guns could make an attack, except on a large scale, slow and costly. Working parties were sent up each night to prepare a new line behind Three Bushes Hill on Sangar Ridge and Umbrella Hill. This line was occupied on the 21st; the Devons then evacuated Three Bushes and passed through the regiment to reserve. Thus ended the battle of Berukin. That it failed to achieve its object did not reflect in any way upon the gallant troops who took part in it. After the initial surprise the enemy put up an unexpectedly stiff resistance, and counter-attacked with the utmost determination. This may have been partly due to the presence of seasoned German troops, with machine-gun and trench-mortar detachments. The German artillery was particularly effective. It is also said that, on the first day of the battle, maps and operation orders fell into the hands of the enemy, which gave away our plans, and enabled him to anticipate them. Even as it was, we came very near to success. A German officer who was present, was of opinion that, had Arara been consolidated on the 10th, the whole position would have become untenable. 'There is no doubt', he concludes, 'that if the attack had been made with more brigades and more artillery support, and had fallen on Turkish troops only, the catastrophic break through would have taken place in April.'

As a matter of fact, we can now see that in the long run the failure actually benefited rather than harmed us. Had our cavalry reached its limited objective at Tulkaram, it would have undoubtedly caused the withdrawal of the Turkish army, with considerable loss, to a line further north, and by this means they would probably have escaped the great enveloping movement which led to their complete annihilation six months later.

Appendix

HONOURS WON AT THREE BUSHES HILL

Military Cross:

Captain R. P. T. Ffrench. On April 11th, 1918 at Three Bushes Hill, he led his company to the attack at dawn with conspicuous gallantry, and though wounded in two places by a bomb, continued to fight on and to inspire his men by his courageous example. During the night April 11th-12th, when the Turks repeatedly heavily attacked, though wounded in the morning, he splendidly conducted the defence of his portion of the line, and greatly inspired the men.

SUBEDAR HARNATH SINGH, I.O.M.[1] During the attack at dawn on April 11th, 1918 on Three Bushes Hill, when officers and men were falling fast, and he was the only officer left in that portion of the line, he assumed command of a platoon, rallied the men and three times gallantly led them against the enemy, his utter disregard for his personal safety greatly inspiring the men.

Indian Order of Merit:

JEMADAR NANIG RAM.[2] On April 11th, 1918 during the attack on Three Bushes Hill, he gallantly led a charge against the Turks, and drove them from their positions. He behaved most gallantly, regardless of his own personal safety, on the night April 11th-12th. When Major Hunt was commanding the line and all other Indian Officers had become casualties, he assumed command of 'B' Company, and in a critical situation when the enemy in force had approached to bombing distance, he held up their counter-attack and drove the enemy off with rifle-fire and bombs, rallying the men and keeping the line intact.

Indian Distinguished Service Medal:

2903. RFM. TOTA RAM. On the morning of April 11th, 1918, during the attack on Three Bushes Hill, he kept his Lewis gun in action under heavy rifle fire, and later on, when almost surrounded, he brought his gun safely out of action. He was wounded later on.

[1] The decoration was presented to him at a Parade at Mhow by Major-General Sir John Shea, K.C.M.G., C.B., D.S.O., Commanding C.P. District.

[2] Killed at the Battle of Megiddo, September 18th.

THE GREAT WAR: MARCH 1918 TO THE DISBANDING OF THE E.E.F.

Chapter XIII

THE GREAT WAR: MARCH 1918 TO THE DISBANDING OF THE E.E.F. (DECEMBER 1920)

While Briton and Turk were struggling for mastery on the banks of the Wadi Ballut, an unparalleled disaster had overtaken our arms upon the Western Front. In March, the great German offensive had virtually annihilated our Fifth Army on the Somme. Scarcely had this onslaught been brought to a standstill, when a second offensive on an equally extensive scale was launched by the enemy further north, threatening to cut us off from our allies, drive us back to the sea and seize the Channel ports. Our casualties had reached the appalling figure of nearly a quarter of a million. The War Office immediately informed General Allenby that he would have to assume the defensive, while the greater part of his army was withdrawn to fill the gaps on the Western Front, and was replaced by Indian and foreign units. The 52nd and 74th Divisions, together with twenty-four battalions of infantry, nine of yeomanry, five and a half batteries of siege-guns and five companies of machine gunners, were despatched to France before the end of July. Sixty British Infantry battalions out of ninety-one were thus sent away, and these were replaced by Indian troops, partly from Mesopotamia and partly new divisions recently recruited, and by French, Italian, Armenian and Jewish contingents. The work imposed upon the staff by this wholesale reorganization almost surpasses the imagination. Many of the new levies consisted of raw recruits who had not even fired their musketry course, and were commanded by officers unable to speak more than a few words of Urdu! Instead of an army of seasoned men who had fought shoulder to shoulder from Gaza to Jerusalem, Lord Allenby and his staff had to make the best of a cosmopolitan mass of troops, of a bewildering variety of language and nationalities. That he could, under these circumstances, win one of the decisive victories of history, stamps him as one of the greatest of military leaders. The time factor was also important. The rains break in Palestine in November, and it was necessary that the blow should be struck before then; the previous year's experience had taught him that campaigning during the wet season in Palestine is fraught with every kind of difficulty. It was therefore necessary for the new army to be ready to strike early in September. From July to September, intensive training went on without a moment's cessation. It was fortunate that the Turk did not attack during the period of transition. But the Turkish armies were in a far worse state than our own. The stubborn Turkish infantry soldier, who had contested every inch of our advance so gallantly, was at last beginning to feel the relentless pressure of the last two years. He had been pushed steadily back, mile after mile, in Palestine and Iraq, and nothing is worse for *morale* than continual retreat. The three great cities most famous in the Muslim world, Bagdad, Jerusalem and Mecca, had been wrested from him. The Arabs were in open revolt against their hated oppressor, and the Bedouin tribes under Lawrence were busy wrecking the trans-Jordan railway and threatening the vital point in the Turkish line of communications at Dera'a. Hungry, ragged and bootless, the enemy still held grimly on; but his powers of endurance had reached the breaking point. Desertions became more and more frequent, and what was even more important, he was beginning to fall out with his German masters, who, he began to suspect, were using Turkey as a catspaw. Nor did the British troops remain inactive during the period. Throughout August extensive raids were carried out, for the purpose of blooding the new troops, and seizing important points of vantage. Some of these were on a very extensive scale, and produced a large haul of prisoners. By the middle of August, Lord Allenby's plans were complete. They were at once simple and comprehensive. His army was composed of the Twenty-first Corps under General Bulfin on the left, from the coast to Ra-fat, the Twentieth Corps under General Chetwode in the centre astride the Nablus road, and Chaytor's force in the Jordan valley on the right. At the attack on the Gaza-Beer-sheeba defences, his plan had been a feint attack on his left,

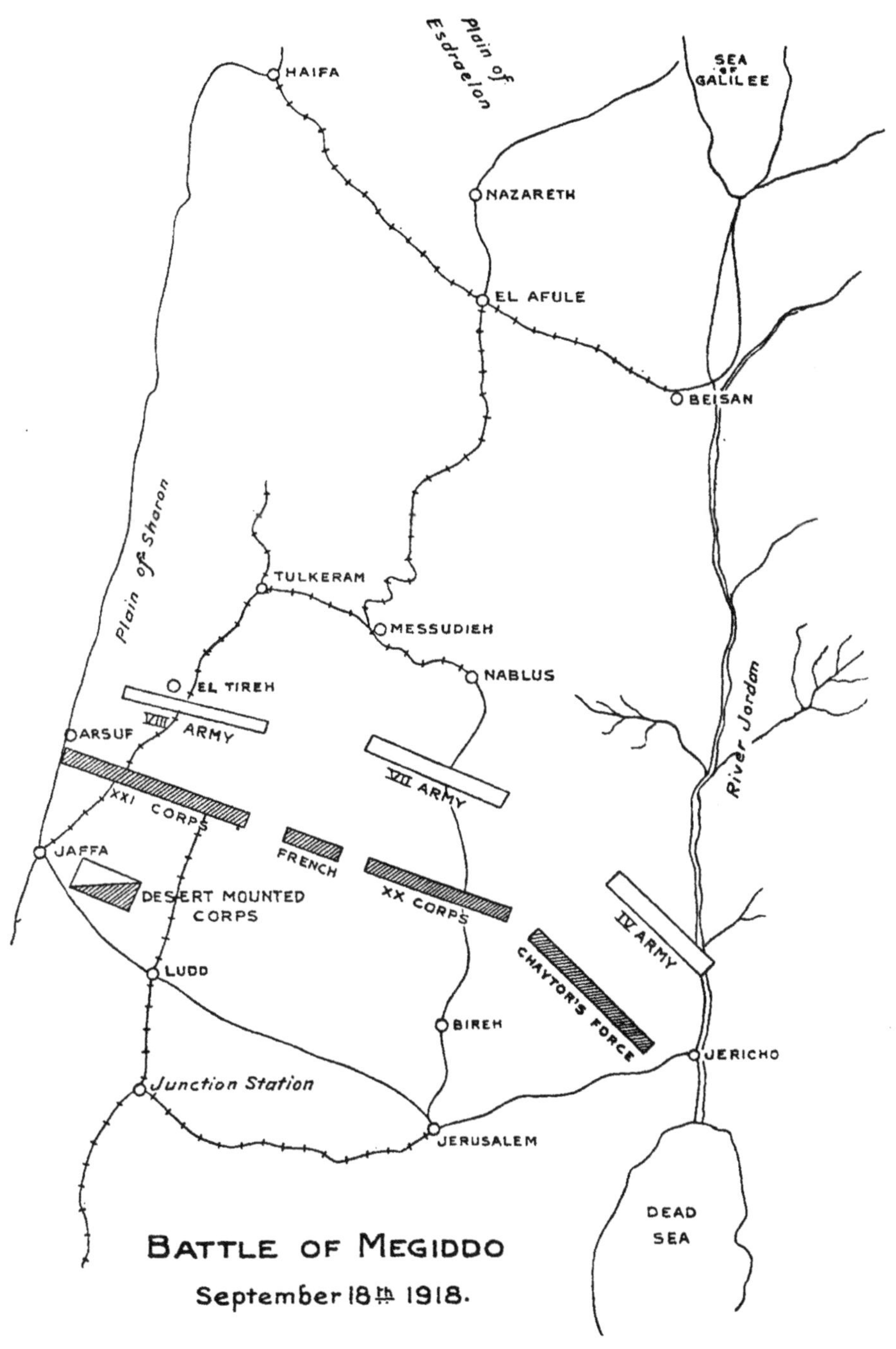

BATTLE OF MEGIDDO

September 18th 1918.

while the cavalry turned the position in his right. Now the reverse was to be the order of the day; the feint was to come on the right and the real attack on the left. Night after night, with the most elaborate precautions and every species of camouflage that experience or ingenuity could suggest, troops were transferred to the left flank, and hidden in the orange groves round Ludd, Ramleh and Jaffa, while behind them the cavalry were massed. Meanwhile the British aircraft effectively prevented any observation of these movements from the air. The rôle assigned to the Twenty-first Corps was to swing round like a gigantic door, the hinge of which was at Ra-fat, and its knob on the coast; as it opened, the cavalry were to pour through the gap, and by a wide sweep, to ride at full speed across the plain of Esdraelon, through the Musmus Pass, and finally to strike at El Fule and Nazareth. Nazareth was the Head-quarters of Liman Von Sanders, and it was even hoped that he and his staff might be surprised and captured. Zero hour was fixed for 4.30 a.m. on September 19th. It was to be preceded by an intense bombardment of fifteen minutes. The order of battle for the Twenty-first Corps was, from left to right, the 60th Division, the 7th (Meerut) Division, the 75th Division, the French contingent and the 54th Division. The objective of the 75th Division was the El Tireh system of defences. This was a formidable obstacle, defended by a network of heavily wired trenches, rifle-pits and machine-gun nests, and what was almost as formidable an obstacle, dense masses of cactus hedges. The order of battle was the 234th Brigade on the left, the 233rd Brigade on the right, and the 232nd Brigade in reserve. The 234th Brigade, under its old commander, Brigadier-General C. A. H. Maclean, D.S.O., now consisted of four battalions, the 1/4th Duke of Cornwall's Light Infantry, the 58th Vaughan's Rifles, the 123rd Outram's Rifles, and the 1/152nd Indian Infantry.

We must now turn back to follow the fortunes of Outram's Rifles. After their relief by the Hampshires on May 8th, they remained for some time in the Berukin sector, in divisional reserve, and were engaged in road-making and other similar occupations. 'C' Company, under Major Kidd, together with Lieutenant C. M. P. Durnford and Lieutenant G. P. Darwell, left to become the nucleus of a newly-raised battalion, the 3/153rd Rifles, and Major Hunt took command, Lieut.-Colonel Cassels having gone on leave to England on May 21st. He returned on July 20th, and after officiating in command of the Brigade for a month, resumed command of the Regiment. August found them in the village of Ra-fat, the scene of much bitter fighting in the battle of Berukin. It was an unhealthy spot, and the men suffered badly from malaria. No one, therefore, was sorry when on September 2nd orders came for a move to the training area at Beit Nabala, where ten days' intensive training was carried out, the troops being thoroughly practised in the formations for the coming attack. On September 17th the battalion marched after dark to the assembly area at Mulebbis. The greatest precautions were adopted to prevent the enemy from observing any movement. No fires were to be lit or lights shown, and all cooking was to be done by means of solidified alcohol. During the day the troops were kept concealed in the Mulebbis woods.

September 18th dawned, warm and cloudless. At 7.30 p.m. the troops were given a hot meal. Waterbottles were filled and they moved out silently to the assembly area, all transport being left in rear. At 3 a.m. by the light of the setting moon, the Battalion took up its position in front of the wire, deploying on the tapes which had been laid down for the purpose by the Staff. It formed up in rear of the centre of the assaulting battalions, the 58th and 1/152nd. The enemy was evidently uneasy and suspicious. Sniping was incessant, varied by occasional shell and machine-gun fire, during those long hours of waiting. The anxiety of those responsible must have been intense. Had the enemy, after all, detected our concentration on his right flank or had it been revealed by deserters? Had he strongly reinforced it, or had he retreated, leaving merely a rearguard to oppose us? But there was no going back now, and at 4.30 a.m., without any warning, our bombardment roared out. Three hundred and eighty-five guns of every calibre, howitzers, trench-mortars and machine-guns, opened rapid fire on the Turkish lines. The destroyers on the coast joined in, and soon the enemy's trenches were enveloped in clouds of smoke and dust. After fifteen minutes, the heavy guns turned their attention to searching more distant points behind the lines, while the field artillery laid down a creeping barrage close behind which the infantry moved forward. The troops were in two waves of two lines each, with 150 yards between each wave, the leading companies in line, and the remainder in artillery formation ; the frontage of each battalion was 400 yards. At first the enemy put down a heavy barrage, but our men swept through it unchecked. In a *wadi* in front of the village of Miske, the 1/152nd found a battery of three howitzers and seven field guns, which they captured at the point of the bayonet. Small bodies of infantry were taken and sent back, but the majority had fled in disorder towards Tulkaram. At 6.45 a.m. all our objectives had been gained, and the cavalry had passed through the lines and taken up the pursuit. The casualties suffered by the battalion during the advance were severe : one Indian officer (the gallant Jemadar Nanig Ram, I.O.M.) and sixteen other ranks killed, and Subedar Major Bhura Ram, I.O.M., and forty-three other ranks wounded. After this, the 75th Division passed into Corps reserve, and the fighting as far as the 123rd was concerned, was over.[1] Meanwhile, the divisions on the left were pushing on relentlessly. Tulkaram, their chief objective, was captured at 4 p.m. by the 60th Division, which had covered eighteen miles, mostly through soft sand, during the course of the day. Meanwhile the cavalry had started on their great ride through the plain of Esdraelon, and at dawn they were actually at the gates of Nazareth, where they narrowly missed capturing Liman Von Sanders himself ; by the same evening they were at Afule. All lines of retreat for the Turkish Army west of the Jordan were now sealed, and every road running west, north and east was picketed. The Twentieth Corps now came into action, driving their opponents northwards into the trap. Meanwhile the aeroplanes were taking terrible toll of the retreating columns, bombing them until the roads were choked with wrecked transport and abandoned guns. By the 21st

[1] The British Officers who took part in the battle were Lieut.-Colonel G. R. Cassels, Captain R. P. T. Ffrench, Lieutenant J. R. Birchall, Lieutenant W. A. L. James, Lieutenant R. D. Ambrose, Lieutenant W. B. Huntley, Lieutenant Tregaskis, Lieutenant H. S. Peppin, and Lieutenant B. H. Kamakaka, I.M.S.

organized resistance had ceased, and the Turks were a disordered mob, frantically endeavouring to escape from the net which encircled them.

Five days later, the victorious troops received a congratulatory letter from the Commander-in-Chief, which aptly summarized the greatness of their achievement. 'I desire to convey to all ranks', wrote Lord Allenby, 'and all arms of the force under my command, my admiration and thanks for their great deeds in the past week, and my appreciation of their gallantry and determination which have resulted in the total destruction of the Seventh and Eighth Turkish Armies opposed to us. Such a complete victory has seldom been known in all the history of war.'

Outram's Rifles, however, were not destined to play a part in these operations. They were ordered to return to the old Turkish line, and assist in salvage work on the trenches, where a large quantity of artillery, machine-guns, rifles and stores was collected and dumped. On October 10th they marched to Afule, where they helped to guard a large prisoners-of-war camp; thousands of Turks and Germans were passed through it on their way to Lejjun. After the Armistice they were moved to Kantara. Everyone was pleased to hear the news in January 1919 that the Distinguished Service Order had been conferred on Lieut.-Colonel G. R. Cassels,[1] and Lieut.-Colonel B. G. B. Kidd, and the Military Cross on Lieutenant B. Kamakaka, I.M.S., but the Battalion suffered a terrible bereavement in the following month when Captain C. F. F. Moore died, on February 18th, of influenza. By his tragic death Outram's Rifles lost a gallant and talented officer, greatly beloved of all ranks. In April, on their transfer to the 31st Brigade, 10th Division, they received the following letter from the Divisional Commander:—

'Before any more units leave the Division and before further demobilization takes place, I wish to thank you all, commanders, staffs and all ranks, for what you have done and for the loyal support you have always given me.

[1] Lieut.-Colonel Cassels was also awarded the Croix-de-Guerre and the Order of the Nile, third class. A summary of the awards won by the Regiment is given in Appendix V at the end of the volume.

'The 75th Division was formed a little over a year and seven months ago. The period is not long, but has been quite long enough for the Division to make a name for itself, and to create the splendid reputation it now holds. The successful raids on the Old British Trenches, Outpost Hill and Middlesex Hill, the break-out from Gaza, the rapid advance to Junction Station, and thence, driving the Turks before it, to Bab el Wad, the capture of Nebi Samwil, the Key to Jerusalem—such is the record of the 75th Division for the first six months of its existence. A few weeks later the Division captured the line Kibbeah-Kh-Ibbaneh-Horse Shoe, followed shortly by the five mile jump to Berukin, El Kefr—Dai Ballut—Um-Taway, and finally after a few months' wearisome trench-warfare, the monotony of which was somewhat lessened by occasional raids, the Division took part in one of the greatest and most successful operations of the War—the total defeat of the Turks and the capture of Palestine. This is a magnificent record, and one of which the 75th Division may be justly proud. It has been attained by courage, good discipline and a fine spirit of comradeship and unselfishness—qualities which go to make the finest soldiers. It is such men I have had the privilege and good fortune to command.

'As we shall soon be separated and scattered to different parts of the world, I take the opportunity of wishing everyone who is in the Division and those who have already left it, a joyful reunion on their return to their homes, and the best of luck and happiness in the future.'

P. C. PALIN,
Major-General,
Commanding 75th Division.'

Head-quarters 75th Division,
March 2nd, 1919.

The remainder of the Battalion's stay in Egypt was uneventful. They were employed in routine duties, guarding wireless stations and bridges and the prisoners-of-war camp. During most of the time they were at Heliopolis. In April 1920, Lieut.-Colonel Cassels left the regiment to take over the command of the 31st Infantry Brigade, and was succeeded by Major R. V. Hunt. Outram's Rifles, during this year, maintained its old reputation for athletics by winning the E.E.F. Hockey Tournament. At last came the welcome order to move to Suez for immediate return to India. The Regiment embarked on the *Orotava* on December 16th, arriving in Bombay on New Year's Eve, after an absence of very nearly four years on Active Service. Prior to leaving Major Hunt received the following farewell letter from the Commander-in-Chief :—

'General Head-quarters,
Egyptian Expeditionary Force,
November 30th, 1920.

To—
The Officer Commanding,
123rd Outram's Rifles.

'On your departure from Egypt and the Egyptian Expeditionary Force, please express to all ranks my high appreciation of the services they have rendered, and their admirable spirit and conduct in all circumstances.

'Your Battalion has worthily upheld the fighting traditions of the Indian Army.

'I thank you, and wish you good luck.

ALLENBY,
Field-Marshal,
Commander-in-Chief,
Egyptian Expeditionary Force.'

THE 3/153rd RIFLES, 1918-22

Lieut.-Colonel B. G. B. Kidd, D.S.O.

[*face p.* 173

CHAPTER XIV

THE 3/153RD RIFLES 1918-1922

IT was felt that the record of Outram's Rifles would be incomplete without some account of the brief but glorious career of the 3/153rd Rifles, as, though this battalion, which was raised at Surafend, near Ludd, on May 24th, 1918 (Empire Day) consisted of units from the 123rd Rifles, 125th Rifles, 105th Mahratta Light Infantry and 124th Baluchis, its Commanding Officer, Adjutant, Quartermaster, and fifty per cent of the men came from Outram's Rifles.

The following British officers joined at Surafend :—

Lieut.-Colonel B. G. B. Kidd	1/123rd Outram's Rifles.
Captain C. M. P. Durnford	1/123rd Outram's Rifles, Adjutant.
Lieutenant G. P. Darwell ..	1/123rd Outram's Rifles, Quartermaster.
Captain H. B. Graveston ..	105th Mahratta Light Infantry.
A/Captain H. V. S. Page ..	125th Napier's Rifles.
Lieutenant E. C. Priestley ..	105th Mahratta Light Infantry.
Lieutenant S. M. M. Craig ..	125th Napier's Rifles.
Lieutenant J. N. Rai, I.M.S.	Medical Officer.

The class composition was :

Punjabi Mohammadans ..	8 platoons.
Jats	2 platoons.
Deccani Mahrattas ..	2 platoons.
Konkani Mahrattas ..	1 platoon.
Hindustani Mohammadans	2 platoons.
Deccani Mohammadans ..	1 platoon.[1]

[1] Subsequently reorganized as follows : 'A' and 'B' Companies, three platoons Punjabi Mussalmans and one platoon Rajputana Jats; 'C' Company, one platoon Punjabi Mussalmans, two platoons Mahrattas, one platoon Deccani Mussalmans ; 'D' Company, two platoons Hindustani Mussalmans, one platoon Punjabi Mussalmans, one platoon Mahrattas.

On June 7th the Battalion marched up to position in the second line, having been appointed to the 158th Brigade in the 53rd (Welsh) Division. It arrived at Taiyebeh on June 10th and was inspected by Brigadier-General H. Vernon, D.S.O. On June 17th, Brigadier-General Pearson, acting in command of the 53rd Division, congratulated the Battalion on the excellent *sangars* it had made in the second line. On June 22nd the Company from the 2/124th Baluchistan Infantry arrived with Captain Mountain, M.C., and Lieutenant Rust. On July 7th Major T. O. Wilkinson, 91st Punjabis, arrived as Second in Command. During the period, on the Nablus Road, July 7th-19th, the Battalion was continuously employed on construction of defences in the first and second lines, and in common with other units in the Brigade received the Corps Commander's thanks for the good work carried out under trying conditions.

On the night of July 20th-21st the Battalion was relieved by the 2/127th Baluchis and marched to Jutna (seven miles) and next night to Ranger corner (ten miles), near Der Ibzia, where the 158th Brigade went into Corps Reserve. The next three weeks were spent in washing, bathing, disinfecting and completing inoculations, sports and competitions. The Regimental Schools of Instruction for young N.C.O.s and for bayonet fighters were kept hard at work. On August 4th, G.O.C. 53rd Division inspected 'A' Company in the field service marching order and expressed himself as very pleased with the turn-out. An interesting Field Day was held on August 15th with one section of the Machine Gun Company and one section of the Light Trench Mortar Battery. The necessity for intercommunication and interchange of runners was emphasized. For the first time the Battalion practised Forward Lewis Guns. The inter-company Rifle Exercise Competition was won by 'A' Company (Jat Platoon) on August 16th.

On August 17th an advance party consisting of four Company Commanders, three Indian officers and three non-commissioned officers per company left for the front line to make preparation for the Regiment to take over from the 1/50th Kumaon Rifles between Wadi Sahlat and Jiljilia.

On August 19th the Battalion marched eleven miles by night to bivouac south-east of Atara. On the night of August 20th-21st it relieved the 1/50th Kumaon Rifles in the line under Lieut.-Colonel Lang at Aliuta.

On August 22nd a severe influenza epidemic commenced and two platoons were borrowed from 3/154th Infantry to enable the Battalion to carry out our duties. These platoons remained until August 31st. During this period there were no operations except for the usual nightly patrols and artillery work on each side. Grapes and figs were in abundance and greatly enjoyed by all ranks. On September 5th the influenza epidemic abated and two platoons from the 3/150th Infantry and the four Lewis gun teams from the 5/6th R.W.F. returned to their units. On September 6th Captain R. G. Mountain temporarily took over the duties of Adjutant vice Captain Durnford, proceeding on short leave. Captain H. P. Rudolf, R.A.M.C., replaced Captain M. A. MacDonald, R.A.M.C., who returned to his unit, the 113th Field Ambulance. Captain MacDonald had been unremitting in his zeal during the influenza epidemic.

On September 7th a fighting patrol under Lieutenant Godfrey, consisting of No. 13 platoon under Subedar Ram Chander Rao and No. 14 platoon under Subedar Sarfaraz Khan, laid up for a Turkish patrol in Wadi Gharis from 7 o'clock to mid-night. No enemy patrol was seen. The patrol also proceeded to scupper an enemy listening post, but the listening post was not held. Our patrol came back under light rifle fire and a few bombs were thrown by the enemy, but we sustained no casualties.

On September 11th a draft of forty-one Jats and thirty-eight Punjab Musalmans arrived from the Indian Base Depot, Kantara; there were no trained bombers or Lewis gunners amongst them, and on September 15th another draft of four Havildars and forty-seven Sepoys arrived from Kantara.

During these days preparations were being secretly made for the great advance, and on September 15th all Commanding Officers assembled at Divisional Head-quarters to meet the Commander-in-Chief, who informed them that he believed there was no way of escape from destruction for the Turkish Armies and that we had 10,000 Cavalry on our left flank, ready to break through.

On September 15th the Battalion was relieved in No. 5 sub-sector by two companies 3/154th Infantry under Lieut.-Colonel Dawson. The Battalion then marched by companies via the Nablus road to Nimr Wadi, where they remained in hiding under the olive trees. Captain Graveston and twenty men remained behind in charge of bivouacs and tents which were all left standing. Lieutenant Rust was temporarily attached to Brigade Head-quarters.

On September 16th the undermentioned British officers were with the Battalion for the operations :—

Commandant	Lieut.-Colonel B. G. B. Kidd.
Acting Adjutant	Captain R. G. Mountain.
Signalling Officer	Lieutenant S. M. Craig.
Intelligence Officer ..	Lieutenant Lunt Roberts.
'D' Company Commander	Captain C. E. W. Reith.
'A' Company Commander	Lieutenant R. E. Godfrey.
'B' Company Commander	Lieutenant H. A. M. D'Este.
'C' Company Commander	Lieutenant F. Loveland.
Medical Officer	Captain H. P. Rudolf, R.A.M.C.
Quartermaster with 2nd line	Lieutenant G. P. Darwell.

On September 16th the Battalion remained hidden in the Wadi Nimr during the day, the Commanding Officer alone going ahead to prepare taking over the line from the 1/7th R.W.F. At 19.30 hours the Battalion marched by Tay Wadi and Tinto Hill and took over No. 2 Sub-sector of No. 1 section from the 1/7th R.W.F. under Major Parkes, coming under the orders of the 159th Brigade.

'A' Company under Lieutenant Godfrey relieved Fusilier Ridge.
'B' Company under Lieutenant D'Este relieved Ide Hill.
'C' Company under Lieutenant Loveland relieved Round Hill.
'D' Company under Captain Reith relieved Abu Felah,
with 2 platoons in Reserve at Battalion Head-quarters.

The relief was completed by midnight.

Major T. O. Wilkinson and sixty first reinforcements returned from Wadi Nimr to the Divisional Dump.

At 15.00 hours on September 18th, Lieut.-Colonel Borthwick, D.S.O., 5/6 R.W.F., visited the line and brought orders from the 159th Brigade for one company to attack Fife Knoll that night. 'B' Company was detailed for this.

At 17.30 hours further orders were received for two platoons to demonstrate against Kh. Amurieh, so two platoons of 'A' Company under Lieutenant Godfrey were detailed. These impending operations necessitated changes in the line; Captain Reith moved to Ide Hill to take over the line from 'B' Company, and two platoons of 'C' Company moved over to Ide Hill to act as support.

At 22.30 on September 18th Lieutenant H. A. M. D'Este and 'B' Company moved out from Ide Hill under cover of trench mortar bombardment, most of the shells of which unfortunately fell short. The Company advanced against the position, meeting with heavy rifle and machine-gun fire and considerable opposition. The Company approached within twenty yards of the enemy position, but were bombed back from the front, the Turks putting down a well regulated barrage by word of command, the enemy at the same time counter-attacking on both flanks. Lieutenant D'Este was wounded at this point and the line withdrew somewhat, but was kept well together by Subedar Mohammad Alam, who was eventually awarded the I.O.M. Considerable casualties (about forty) had been suffered at this time.

At 00.15, two platoons of 'C' Company, under Captain Reith, were ordered up to reinforce 'B' Company and take the Knoll, and Lieutenant Craig was sent from Battalion Head-quarters to take command at Ide Hill. 'B' Company with the support of these two platoons again arrived within twenty yards of the enemy's *sangars*, but were forced to retire by heavy machine-gun and rifle fire and bomb barrages put down by the enemy.

At 01.25 'B' Company and two platoons 'C' Company then withdrew on to Ide Hill. Although the knoll was not taken, the object of the attack was accomplished, as a large number of enemy reinforcements were concentrated on Fife Knoll, thus allowing the troops on our right to break through.

The casualties were heavy, and were as follows :—

'B' Company. Lieutenant H. A. M. D'Este wounded.
Killed, 2 I.O.Rs.
Missing, believed killed, 2 I.O.Rs.
Wounded, 52 I.O.Rs.
'C' Company. 1 I.O.R. missing, believed killed.
1 I.O.R. wounded.

Captain Reith and Subedar Ali Mohammad Shah were slightly wounded, but remained at duty; the Subedar received an immediate award of the I.D.S.M.

Simultaneously with the attack on Fife Knoll, the fighting patrol under Lieutenant Godfrey was demonstrating against Kh. Amurieh, which it found strongly held. Considerable rifle fire was met with. Subedar Bakhta Ram, 123rd Rifles, was killed, six I.O.Rs. were wounded and one missing, believed killed. The patrol returned to Fusilier Ridge at 01.05.

Throughout the night Ide Hill and Battalion Head-quarters were continually shelled, and severely between 23.00 and 03.00, when the shelling practically stopped.

At 11 o'clock on September 18th the Battalion was relieved by two companies of the 3/151st Infantry under Lieut.-Colonel Doveton and concentrated at the Junction of the Tay and Kola Wadis. Subedar Bakhta Ram was buried on the south side of the Tay Wadi. At about 19.00 hours the Battalion was visited by Brigadier-General H. Vernon and Captain More, Brigade Major, and again came under the 158th Brigade.

At 23.00 o'clock on September 19th the Battalion moved up the Forth Wadi, in which it was distributed for road making, with orders to work till 04.00 on the following day. At 03.40, the work having been completed, the Battalion concentrated and formed up south of Ruin Hill, where the Brigadier told it to remain in a state of readiness. At 08.05 the Battalion advanced up the re-entrant between Hind Head and Ruin Hill in support of the 4/11th Gurkha Rifles, halting at Kulason for attack on Hill 2906. On Hind Head we met Major-General Mott and Lieut.-Colonel Garcia, G.S.O. The 4/11th Gurkhas had instructions to advance at 09.15 but not to get heavily engaged. At 11.40 'A' and 'D' companies moved up in support of the

Gurkhas and at 14.15 the whole battalion moved forward to attack Hill 2906, one mile west of Domeh, in conjunction with the 4/11th Gurkhas on our left. At 14.45 Lieutenant Loveland commanding 'C' company, was wounded, and Lieutenant Lunt Roberts took over command of the company. The attack proceeded up a ridge, under some machine-gun fire north of Wadi Sebbas. Here the 4/11th were hung up. Just as the attack was re-starting orders were received by telephone from Brigade Head-quarters to stand fast until an artillery barrage was put down at 15.55; but the psychological moment was then lost. The artillery barrage was eventually cancelled and we prepared for a night attack and got the guns of a Field Battery ranged. At dusk some re-organization took place; the 4/11th Gurkha Rifles took over the left of the line and the 3/153rd the right, 'D,' 'A' and 'C' companies being in the line and 'B' company in close support. These were ready to attack on receipt of orders from the Brigade. The objective, however, was abandoned by the enemy and the 4/11th Gurkha Rifles moving forward occupied the position, notifying at 04.30 by telephone that this had been effected. On receipt of this news, in accordance with orders, the Battalion concentrated and got on to a roadway near its left flank, just south of El Kust. Here the 1/153rd Infantry under Lieut.-Colonel Withers were on the march, and shortly afterwards, about 05.15, Major-General Mott passed through. The night had been cold, with a thick, damp mist.

Marching north, the Battalion rendezvoused at Brigade Head-quarters near Nejmeh at 06.30, the 1/153rd moved on westward and halted about a mile away, the 3/154th Infantry under Lieut.-Colonel Dawson were halted just near Brigade Head-quarters, where were Brigadier-General H. Vernon, D.S.O., commanding 158th Brigade, and Captain More, Brigade Major.

The casualties during the previous days' advance had been:—

1 British Officer wounded (Lieutenant Loveland).
'A' Company. Nil.
'B' Company. Wounded, 1.
'C' Company. Wounded, 3; killed, 1.
'D' Company. Wounded, 16; killed, 2.

At 5.30 a.m. Captain Durnford rejoined the Battalion from leave, having had great difficulty in getting up from the Corps Reinforcement Camp. He took over the duties of Adjutant from Captain Mountain (who had been answering for him) at 7 a.m.

At 8 a.m. the Battalion was ordered to be in readiness to support the 3/154th Infantry in their attack on Akrabeh. At 10.15 the 3/154th had occupied the objective and driven off the enemy. It was then ordered to be ready to continue the attack and capture Yanun and height El Tuwanik (2,854 feet), two miles to the north of it.

At 11.05 it commenced the advance in artillery formation over a mile of open plain, coming under light shell fire when near Akrabeh, but the only casualty was one mess donkey killed by shrapnel and three mules wounded.

After Akrabeh it entered the hills, captured Yanun, where it left a lot of booty, and advanced up El Tuwanik in the hottest part of the day. Major Windsor, R.F.A., accompanied the advance. The only opposition was caused by a few snipers who were captured. The top of the hill was reached at 15.50 hours, the men by then being in a very exhausted condition. From the summit Nablus could be seen some ten miles away to the west, and in the valley immediately to the north, Turkish transport was seen moving away north and north-east, being bombed by our aeroplanes.

At 17.10 orders were received to continue the advance and occupy the village of Beit Dejan, about 4,000 yards as the crows flies, but approximately double that distance by the zig-zag route the Battalion had to follow down the mountains. Beit Dejan was on the edge of an olive wood, in which camp fires could be seen burning ; in the wood the advance guard under Captain Reith captured some dozens of Turkish prisoners. The village was occupied at 22.15 and the roads of escape for the Turks towards the East were closed, the 4/11th Gurkha Rifles, under Lieut.-Colonel W. L. Dundas, being in position near by at Tana Ruins. The number of prisoners was increased to forty-two and three machine-guns were captured, also considerable quantities of stores, equipment and ammunition.

Water was found at Beit Dejan in sufficient quantities for the men.

The total advance since eight in the morning was about 18,000 yards direct over very difficult country. No one fell out during the day, though the sun was very hot during the advance on El Tuwanik and the ascent of the last 1,000 feet was very steep ; the men began to feel the lack of water.

At 08.00 on September 22nd orders were received to remain at Beit Dejan during the day. The surrounding heights were picketed and Lieutenant Lunt Roberts, the intelligence officer, prepared a fine panoramic view of the country to the north.

The Battalion collected considerable quantities of enemy material into dumps at the roadside and vast quantities of enemy documents and maps were sent in to General Head-quarters.

At 23.00 orders were received to withdraw from Beit Dejan on the morrow to Yanun, leaving behind two platoons.

On September 23rd at 10.00, leaving Lieutenant Godfrey and two platoons at Beit Dejan, the Battalion marched back to Yanun, where further quantities of enemy material were collected. At 15.00 hours at Yanun orders were received to withdraw Lieutenant Godfrey and party, who rejoined the unit at 22.00 hours. The afternoon at Yanun was busily occupied in sending recommendations for awards, returns and re-organizing generally. Next morning on September 24th at 9 a.m. the Battalion set off via the Wadi Jenab to the Nablus road and bivouaced at Yetma, where it was generally believed that it would remain about six weeks in order to remake the Nablus road. Complimentary letters were received from the Brigade and Divisional Commanders.

On September 25th under the directions of Lieut.-Colonel Watson of the Pioneers, work was commenced on the Nablus Road. Lieutenant Rust rejoined from Brigade Head-quarters.

The following is the summary of the casualties during the operations commencing 17/9/18.

	B.Os.	I.Os.	I.O.Rs.
Killed ..	–	1	8
Wounded ..	3	1	82

Orders were received to pack up and move off the next day and on September 26th the Battalion marched in Brigade southwards along the Nablus road to the neighbourhood of El Tell, where it bivouaced in Wadi Jib.

On September 28th Major T. C. Wilkinson, Captain H. B. Graveston, Subedar Major Mubarik Ali Khan, Jemedar Mohammad Nur and twenty-eight I.O.R.s rejoined from first Reinforcement Camp. On September 28th Brigadier-General H. Vernon left the Brigade, to its great regret; the command was temporarily taken by Lieut.-Colonel Borthwick, 5/6th R.W.F.

On September 29th the Battalion was inspected by Major-General T. Mott, Commanding 53rd Division, who expressed himself as very pleased with what the Battalion had done in the recent operations.

The Battalion remained near El Tell until October 10th, when it marched to Ramleh via Der Ibzia and Latron.

The first march was a very trying one with several very steep and long ascents. No one fell out. On the next day en route to Latrun one man fell out, and none the third day, when Ramleh was reached. The Regiment had fewer men falling out than any regiment in the Division; large numbers fell out in all other units. The Regiment was congratulated by the Brigade Commander and Divisional Staff on its march discipline.

The marching in state of the Regiment at Ramleh on October 12th, 1918 was :—

Lieut.-Colonel B. G. B. Kidd	Commandant.
Major T. O. Wilkinson ..	2nd-in-Command.
A/Captain C. M. P. Durnford	Adjutant.
Lieutenant G. P. Darwell ..	Quartermaster.
Lieutenant S. M. Craig ..	Signalling Officer.
Lieutenant N. A. Rust ..	Intelligence Officer.
Lieutenant R. E. Godfrey ..	Commanding 'A' Company.
A/Captain R. G. Mountain	Commanding 'B' Company.
A/Captain H. B. Graveston	Commanding 'C' Company.
A/Captain C. E. W. Reith ..	Commanding 'D' Company.
A/Captain H. P. Rudolf, R.A.M.C.	Medical Officer.

The Regiment remained at Ramleh until November 2nd, where it entrained for Alexandria, arriving on November 4th.

Lieut.-Colonel B. G. B. Kidd was mentioned in Sir E. H. E. Allenby's despatch of October 23rd, 1918.

On October 31st, General Sir Edmund Allenby, G.C.B., etc., Commander-in-Chief, Egyptian Expeditionary Force, was good enough to accept an invitation to attend a 'tamasha' that evening in the lines to celebrate the signing of the Armistice by Turkey. A Guard of Honour of a hundred rifles under Captain Durnford was provided, which the Commander-in-Chief said was 'remarkably fine'; the camp was lit by hundreds of torches. After the show the Commander-in-Chief made a speech to the Battalion, congratulated them on what they had done, and told them that they had taken part in the total destruction, in a few days, of three Turkish Armies.

On November 7th, 1918, sanction was accorded for the Battalion to be called the 3/153rd Rifles.

On November 21st the Divisional Commander, Major-General Mott, presented medal ribbons to the following, who had been granted immediate awards during the recent operations :—

	Subedar Ali Mohammed Shah.	I.D.S.M.
19.	Havildar Amir Ali.	I.D.S.M.
247.	L/Naik Mohammed Zaman.	I.D.S.M.
25.	Rfm. Shah Mohammed.	I.D.S.M.

On December 1st, 1918 the 53rd Division was inspected at Alexandria by Sir Edmund Allenby. It marched past in Mohammed Ali Square and then through the streets of Alexandria.

On December 11th the Battalion school was increased to eighty and company schools to forty each.

In the *Gazette* of 1.1.19, Lieut.-Colonel B. G. B. Kidd was awarded the D.S.O.

On September 1st, 1919 Jemedars Mohammad and Sayed Pir and twenty-three I.O.Rs. proceeded on tour to Mecca. The King of the Hejaz treated them with extreme courtesy and they were quite pleased with the trip and what they saw.

In the Brigade Assault at Arms on January 13th the Battalion obtained nine firsts and four seconds and at the 53rd Divisional Assault at Arms on January 28th the Battalion obtained four firsts and two seconds.

On February 5th, 1919, an advance party under Lieut.-Colonel B. G. B. Kidd and the Adjutant, Captain Durnford, left for Suez for embarkation to India, it being expected that the rest of the Battalion would shortly follow. As it turned out, however, they were detained in Egypt for nearly a year as part of the Army of Occupation. During March they were on duty in connexion with the protection of the railway line from Arbain Station to Sarapeum, and were constantly employed in patrolling the permanent way and providing guards for the Isolation Hospital and Aerodrome at Port Tewfik. Three Emergency Platoons were kept in constant readiness to move, and at the end of May, 'A' Company, under Lieut. Orpen, was sent to Damietta to quell disturbances there.

On June 28th news was received of the signing of the Peace Treaty with Turkey; the troops at Mansura, headed by the 3/153rd, made a triumphal march, presented arms and fired three rounds of blank. Three cheers were given at the Allies' Consulates and a feast was held in the evening. Subedar Major Mubarik Ali Khan Bahadur, Havildar Major Mani Ram and Rifleman Sita Ram Parab went to England to represent the regiment at the Peace Celebrations.

It was not until January 5th, 1920, that the Battalion embarked for Bombay on the *Kildonian Castle*; it was commanded by Captain H. B. Graveston, Major T. O. Wilkinson having gone on leave. The voyage was marred by an outbreak of influenza, but on the 26th the Battalion marched into Ahmednagar, being played in by the Band of the 119th Infantry. It was met by Lieut.-Colonel B. G. B. Kidd, D.S.O., and representatives of the Depot, and every man was given cigarettes and garlanded. The evening was spent in a feast and nautch.

The following British officers were on the strength at the time:—

Lieut.-Colonel B. G. B. Kidd, D.S.O.
Captain C. M. P. Durnford.
Captain H. B. Graveston.

Captain G. P. Darwell.
Captain R. E. Godfrey.
Captain G. B. Dore.
Lieutenant N. A. Rust.
Lieutenant A. L. R. Duke.
Lieutenant W. H. Price.

In May the Battalion moved to Bombay, where it remained until August 14th, when orders were received to mobilize for Field Service in Mesopotamia, where the Arab Rebellion had broken out. They sailed from Alexandra Dock on September 9th on the *S.S. Cooeyanna*, arriving at Basra on the 18th, after a hot and rough voyage. They at once entrained for Nasiriyah, where they joined the 7th Brigade, under Brigadier-General A. le G. Jacob. They took over the left bank defences, and there was a good deal of sniping, fortunately without much effect.

On November 10th the Battalion joined the 34th Brigade and missed a sharp fight with the Arabs by a few hours: the column was employed in minor punitive operations to the end of the month, when the rebellion came to an end. They were, however, on garrison duty in Mesopotamia for some time longer, and on January 14th, 1922, formed part of the Shatrah Column, which marched to Karaidi and back. It was not until April that they returned to India and were disbanded.

During its brief but distinguished career the British and Indian officers of the Battalion had gained five mentions in despatches: the Commanding Officer had been awarded a D.S.O., an Indian officer the Order of British India, another the Indian Order of Merit, and a third the I.D.S.M. The last-named decoration was also won by three of the rank and file, and five of these, in addition, gained the I.M.S.M.

APPENDIXES

DISTINCTIONS

Distinguished Service Order.	Lieut.-Colonel B. G. B. Kidd.
Mentioned in Despatches.	Major T. O. Wilkinson.
	Lieutenant R. E. Godfrey.
	Subedar Mohammed Hassan.

Order of British India with Title of Bahadur.	Subedar Major Mubarik Ali Khan.
Indian Distinguished Service Medal.	Subedar Ali Mohammed Shah. Havildar Amir Ali. L/Naik Mohammed Zaman. Rfm. Shah Mohammed.
Indian Meritorious Service Medal.	Havildar Kheta Ram. Havildar Per Mohammed. Naik Abdul Karem Ahmedz. L/Naik Babu Jadhav. Rfm. Sarup Ram.
Indian Order of Merit.	Subedar Mohammed Alam.

LIST OF OFFICERS PRESENT WITH THE REGIMENT AND DEPOT, JANUARY 1ST, 1921

REGIMENT IN THE FIELD:

Lieut.-Colonel B. G. B. Kidd, D.S.O.
Captain J. P. Richmond.
Captain G. F. Darwell.
Captain R. E. Godfrey.
Captain G. B. Dore.
Captain S. E. Williams.
Lieutenant R. Chouler.
Lieutenant P. S. Le Geyt.
Lieutenant E. W. B. Gibbs.
Lieutenant A. L. R. Duke.
Lieutenant T. H. Padgham.

Subedar Major Mubarik Ali Khan Bahadur, I.D.S.M., and 15 Indian Officers.

Indian Other Ranks: 771.

AT THE AHMEDNAGAR DEPOT:

Captain H. B. Graveston.
Captain J. A. Greenwood.
Lieutenant H. A. M. D'Este.
Lieutenant H. C. F. Wotherspoon.

Subedar Ghulam Mohammed, I.D.S.M. and 4 Indian Officers.

Indian Other Ranks: 351.

POSTSCRIPT: 1921-2

UNIFORM AND ARMS, 1921

[*face p.* 189

CHAPTER XV

POSTSCRIPT. 1921-2.

January 1921 found Outram's Rifles back in their old quarters in Malcolm Lines, Mhow, where they received a message of welcome from Major-General Sir John Shea, congratulating them on their magnificent war record, and wishing all ranks enjoyment of their well-earned leave. During the ensuing month, the men went on two months' war leave, and those in excess of strength were demobilized. In November, H. E. Lord Rawlinson, the C.-in-C., visited Mhow and inspected the garrison. In a message to Major-General Cassels (now Sir Robert Cassels, G.C.B., etc., G.O.C., Northern Command), he spoke of the parade as 'a pleasure to witness', and he remarked particularly on 'the fine physique and soldierly bearing of the 123rd Outram's Rifles'.

In October 1921, an important measure of reorganization in the Indian Army was carried out. Units of approximately the same composition were grouped together, and numbered according to seniority ; one regiment of each group was formed into a training battalion for the whole group. Each regiment in the group sent one British officer and a certain number of Indian officers and N.C.O's as instructors to the affiliated company in the training battalion, which was then made responsible for recruiting, and the training of recruits and reservists. The group to which Outram's Rifles belonged was now known as THE SIXTH RAJPUTANA RIFLES. It consisted of the following battalions :—

104th Wellesley's Rifles now The 1st Battalion (Wellesley's).

120th P.W.O. Rajputana Infantry now The 2nd Battalion (Prince of Wales' Own).

122nd Rajputana Infantry now The 3rd Battalion.

123rd Outram's Rifles now The 4th Battalion (Outram's).

125th Napier's Rifles now The 5th Battalion (Napier's).

13th Rajputs now The 10th Battalion (Shekhawati).

Numbers 6 to 9 were left blank for expansion in war, and the 10/6th Rajputana Rifles became the Training Battalion.

At a meeting of Commanding Officers held in the following year it was decided that :—

(1) The Group should be called the 6th Rajputana Rifles, the eventual designation of the battalion being 4/6th Rajputana Rifles (Outram's).

(2) Black belts and boots, rifle-green hose-tops, and Rifle mess kit, as in the 60th Rifles, should be worn.

(3) British Officers should wear a rifle-green flash on the left side of the helmet, with 6th R.R. in red.

(4) There should be a single crest, viz. : a Maltese Cross and laurel-wreath, as in the 104th Rifles.

Meanwhile the Regiment was making a name for itself at hockey. In November 1921 it won the Southern Command Tournament at Poona, beating the 3rd Sappers and Miners, after a thrilling struggle by 4 goals to 3. It thus qualified to represent the Southern Command in the All India Hockey Tournament at Meerut. Here in the following March it beat the 46th Punjabis by 3 goals to 1, and carried off the Colonel's Challenge Cup, presented by the Colonel of the Native State Forces. The team which won this epic victory was composed of the following :—

Captain H. S. McEntire.	Lieutenant F. Walton.
Havildar Gulji Singh.	Havildar Jhoki Ram.
Naik Jowaru Ram.	Naik Firoz Khan.
Naik Abdul Majid.	Naik Mora Ram.

L/Naik Gopal Ram.
L/Naik Bhani Ram.
L/Naik Lalchand Ram.

The Regiment also maintained its old reputation for shooting, winning the Prince of Wales' Cup for Lewis Gun Teams, the Commander-in-Chief's Cup, and the Malerkotla Cup in

Lieut.-General Sir Walter Delamain, K.C.B., K.C.M.G., D.S.O.

[face p. 191

1923-4, and the Mother Country and Cawnpore Woollen Mills Cup in 1926-7.

The inspection report of the year 1921-2 spoke of the Battalion as 'well commanded and well trained. Fit for service. A very well administered unit with a high standard of discipline. Turn-out always good. Much interest is always shown in the welfare of the men : games are encouraged on the right lines with very successful results.' The high praise brought a chorus of congratulations from Sir Walter Delamain, Sir Robert Scallon and others. The latter, writing to Lieut.-Colonel Kidd, said : 'It really *is* a splendid report. I don't think the *Paltan* has ever had a better, and I do most heartily congratulate you and the *Paltan* at large.'

In February 1922 Outram's Rifles had the privilege of furnishing the Guard of Honour for H.R.H. the Prince of Wales, on his visit to Indore. The Guard was composed of Jats of 'C' Company under the command of Captain R. P. T. Ffrench, M.C., and was accompanied by the Band. His Royal Highness recalled the fact that by a notable coincidence the same regiment furnished the Guard of Honour for his grandfather on the occasion of his visit to Indore in 1876. He afterwards inspected the garrison and was introduced to the Indian Officers ; and he personally congratulated Lieut.-Colonel Kidd on 'the very fine Guard of Honour provided at Indore and on the fine appearance of the Regiment at the Ceremonial Parade'.

On July 19th all ranks heard with sorrow of the retirement of Lieut.-General Sir Walter Delamain, K.C.B., K.C.M.G., D.S.O., Adjutant General of the Army in India, on account of ill-health, after having served it faithfully and gallantly for forty years. He had his first experience of active service with the Berkshire Regiment in Egypt in 1882, for which he was awarded the Bronze Star. Three years later he was transferred to the Indian Staff Corps, being posted to the 23rd Regiment of Bombay Light Infantry, as officiating Wing Officer, on February 1st, 1885. He served in Burma, 1885-8, for which he was awarded the medal and two clasps ; with the Zeila Field Force in Arabia, 1890 ; with the China Field Force, 1900, where he was mentioned in Despatches ; in the Aden Hinterland,

1903-4, for which he received the D.S.O.; in the Persian Gulf, 1910-11; and lastly in the Great War, where he commanded the Poona Brigade of the 6th Poona Division. As regards the latter, it was truly said that 'where hard knocks were to be given and received, there the Poona Brigade and its commander were surely to be found'. He was in the battle of Shaiba and the subsequent operations, and with Townshend in his advance on Bagdad, and the retreat to Kut-el-Amara, and its siege and capitulation. Ever a true friend of the sepoy, he, together with Sir Charles Melliss, V.C., worked heroically to mitigate the sufferings of the unfortunate garrison while they were prisoners of war in Turkey, and his exertions won him the gratitude of all ranks. After the Armistice, he returned to India as Adjutant General, and his intimate knowledge of the requirements of the rank and file, and his long experience were invaluable in the period of post-war reconstruction. He commanded the 123rd Outram's Rifles from January 1905 to January 1912. He is one of the four great names, *clara et venerabilia nomina*, Outram, Gatacre, Scallon and Delamain, which will always be indissolubly connected in men's minds with Outram's Rifles.

On November 22nd, 1922 the Regiment celebrated its centenary, which had been delayed by the Great War, and unfortunately, it was impossible to hold the celebration as had been planned, on 'Kirkee Day', November 5th. Sports and an Assault at Arms were held in the morning and were witnessed by Major-General and Mrs. Clery and most of the Station. In the evening the Officers entertained General and Mrs. Clery and the ladies of the Regiment to dinner in the Mess, and afterwards the Indian Officers were 'At Home' to the Station at a theatrical performance. Telegrams of congratulation were received from Lieut.-General Sir John Gatacre, General Sir Robert Scallon, and Lieut.-General Sir Walter Delamain. A feature of the celebrations was the presence of many veteran Indian officers, including that grand old soldier, Captain Sheikh Mahbub Sardar Bahadur, who had joined the Regiment over half a century ago, and had won distinctions in the Afghan and Burmese Wars, Subedar Major Jaita Ram, and Subedar Major Bije Singh, I.D.S.M., who had distinguished himself in the Aden Hinterland.[1] This, perhaps, is a fitting place to say farewell to a great and gallant regiment, which for over a century has upheld the honour of the British Raj in the stifling deserts of Upper Sind and the pestilential swamps of Burma, on the blood-stained slopes of Nebi Samwil and the *sangars* of Three Bushes Hill, in the Persian Gulf and the Aden Hinterland, and the innumerable other places to which duty has called them, since their baptism of fire on that far off November day on the Kirkee plain, where, faithful to their salt, they beat off the flower of the Maratha cavalry. It is a record of which any regiment in the world may be proud. Sweeping changes are inevitable in the near future : old landmarks are being swept away on every hand, but a study of its long and gallant career makes it abundantly clear that OUTRAM'S RIFLES will ever worthily uphold the grand ideals of its historic past.

[1] *Vide* pp. 98, 111 and 113.

APPENDIXES

APPENDIXES

Appendix I

THE 6TH RAJPUTANA RIFLES

Battle Honours.—'Mysore,' 'Seringapatam,' 'Bourbon,' 'Kirkee,' 'Beni Boo Alli,' 'Meeanee,' 'Hyderabad,' 'Aliwal,' 'Multan,' 'Punjab,' 'Reshire,' 'Bushire,' 'Koosh-ab,' 'Persia,' 'Central India,' 'Abyssinia,' 'Kandahar 1880,' 'Chitral,' 'Afghanistan 1879-80,' 'Burma 1885-87,' 'British East Africa 1898,' 'China 1900,' 'Afghanistan 1919.'

The Great War. 'Givenchy 1914,' 'Neuve Chapelle,' 'Aubers,' 'Festubert 1915,' 'France and Flanders 1914-15,' 'Egypt 1915,' 'Gazi,' 'Nebi Samwil,' 'Jerusalem,' 'Tell Asur,' 'Megiddo,' 'Sharon,' 'Palestine 1917-18,' 'Basra,' 'Shaiba,' 'Kut-el-Amara 1915-17,' 'Tigris 1916,' 'Baghdad,' 'Mesopotamia 1914-18,' 'Persia 1918,' 'East Africa 1914.'

Colonel-in-Chief. Colonel H.R.H. The Prince of Wales and Duke of Cornwall, K.G., K.T., K.P., G.C.S.I., G.C.M.G., G.C.I.E., G.C.V.O., G.B.E., M.C., P.C. (Personal A.D.C. to the King).

1st Battalion (Wellesley's) late 104th Wellesley's Rifles.
2nd Battalion (Prince of Wales' Own) late 120th P.W.O. Rajputana Infantry.
3rd Battalion (Prince of Wales' Own) late 122nd Rajputana Infantry.
4th Battalion (Outram's) late 123rd Outram's Rifles.
5th Battalion (Napier's) late 125th Napier's Rifles.
(6th-9th Territorial Battalions.)
10th Battalion (Shekhawati) late 13th Rajputs. (Training Battalion.)

4TH BATTALION (OUTRAM'S) (LATE 123RD OUTRAM'S RIFLES)

Class Composition. Rajputana Jats, Rajputana Rajputs and Punjabi Musalmans.

Raised in 1820 to a considerable extent from men who had served at the battle of Kirkee from the disbanded Dapuri Battalion (Peshwa's service), under Major Ford, Madras Army, and was originally designated the 13th Regiment of Bombay Native Infantry. Became the 23rd Regiment of Bombay Native Infantry, 1824, the 23rd Regiment of Bombay Native (Light) Infantry, 1841; the 23rd Regiment of Bombay (Light) Infantry, 1885; unauthorizedly designated the 2nd Battalion the 23rd Bombay Rifles, 1901; 123rd Outram's Rifles 1903; present designation, 1922.

COLONEL:—Brigadier-General G. R. Cassels, C.B., D.S.O.

Appendix II

COLOURS OF THE REGIMENT

The Regiment, prior to the time when it became a Rifle Regiment in 1888, had three sets of Colours, all of which are extant. The earliest (1821-1852) is now in the collection of the late Major S. G. Everitt. The other two hang on the South and North sides of the Chancel of All Saints' Church, Kirkee, and were encased by Sir George Lloyd, now Lord Lloyd, when he was Governor of Bombay. There is a brass tablet beneath each, and another midway between the two on the floor of the Nave, bearing the following inscriptions :—

South Side

Colours of the 23rd Bombay Native Infantry
(Formerly the 12th Native Infantry),
Presented in 1852 by Mrs. Watkins.
They were borne through the following campaigns
1856-8 Persia, Battle of Mohumra
1858-9 Central India, Pursuit of Tantia Topi
1865 Okhamandal
Deposited in this Church 1870

Centre

In Commemoration of the Past History
of the 23rd Regiment Bombay Native Light Infantry
The above Colours are by Permission
Placed in this Church
A.D. 1870-1889

North Side

Colours of the 23rd Bombay Native Infantry
(Now the 123rd Outram's Rifles)
Presented in 1870 by Lady Spencer
They were borne through the following campaigns
1880-1 Afghanistan
1885-8 Burma
Deposited in this Church 1889

Appendix III

OFFICERS ON THE STRENGTH AT THE OUTBREAK OF THE GREAT WAR

Rank.	*Name.*	*Appointment.*	*Remarks.*
Lieut.-Colonel	H. E. C. B. Nepean	Commandant	A.Q.M.G., Simla.
Major	E. E. Bousfield	Second in Command	Killed in France, 1915.
Major	R. W. C. Blair	D.C. Commander	Commandant 2nd Battalion, 1916.
Major	B. G. B. Kidd	D.C. Commander	
Captain	G. E. Hardie	D.C. Commander	
Captain	W. P. M. Sargent	S.S.O., Bhamo	Transferred to 104th Rifles.
Captain	W. Greatwood	D.C. Commander	Killed in Mesopotamia, 1916.
Captain	W. R. Daniell	D.C. Commander	Killed in Palestine, 1917.
Captain	R. V. Hunt	D.C. Officer	
Captain	A. K. Norris	Brigade Major, Mandalay Brigade	
Captain	J. G. Rae	Adjutant	At the Training Depot, Mhow.
Captain	K. B. McKenzie	D.C. Officer	Killed in France, 1915.
Lieutenant	W. Odell	D.C. Officer	Killed in Mesopotomia, 1917.
Lieutenant	R. Tilley	D.C. Officer	Burma Military Police.
Lieutenant	C. F. F. Moore	Quartermaster	Died at Kantara 1919.
Lieutenant	R. P. T. Ffrench	D.C. Officer	
2/Lieutenant	L. A. Stuart	D.C. Officer	

Appendix IV

ROLL OF HONOUR

123rd OUTRAM'S RIFLES

4th August 1914—20th June 1919

Lieut.-Colonel E. E. Bousfield. Died of wounds in France, 25th September 1915.

Major W. Greatwood. Killed in action in Mesopotamia, 8th March 1916.

Major W. R. Daniell. Died of wounds in Palestine, 1st December 1917.

Captain K. B. McKenzie. Killed in action in France, 25th September 1915.

Captain W. Odell, M.C. Killed in action in Mesopotamia, 22nd February 1917.

Captain C. F. F. Moore, M.C. Died in Egypt, 18th February 1919.

Lieutenant C. N. Harris. Killed in action in Mesopotamia, 21st April 1917.

Subedar Jehangir Khan. Died of disease in Mesopotamia, 23rd December 1915.

Subedar Shivnath Singh. Killed in action in Mesopotamia, 9th January 1916.

Subedar Karam Din. Died of wounds in Palestine, 17th November 1917.

Subedar Mohammed Zaman. Killed in action in Palestine, 11th April 1918.

Jemadar Din Mohammed. Killed in action in France, 25th September 1915.

Jemadar Birbal Ram. Killed in action in Mesopotamia, 5th November 1917.

Jemadar Arjun Ram, I.O.M. Died of wounds in Palestine, 11th April 1918.

Jemadar Nanig Ram, I.O.M. Killed in action in Palestine, 19th September 1918.

And 321 non-commissioned officers and men and three followers in various theatres of war.

Erected by British and Indian Officers 123rd Outram's Rifles.

Their Name Liveth for Evermore

[Tablet in All Saints' Church, Kirkee. The name of Subedar Bakhta Ram, killed in action at Fife Knoll, on 18th-19th September 1918, while serving with the 3/153rd Rifles, does not appear on it.]

The War Memorial of the 6th Rajputana Rifles, on which are inscribed the names of all the fallen of the Regiment, stands on the Parade Ground at Nasirabad. It consists of a white marble *chhatri* or shrine, with a perpetually burning lamp suspended from the centre of the dome by a silver chain.

Appendix V

DISTINCTIONS WON IN THE GREAT WAR

Distinguished Service Order:
Lieut.-Colonel G. R. Cassels.
Lieut.-Colonel B. G. B. Kidd.

Order of the British Empire:
Major J. G. Rae.

Member of the British Empire:
Captain G. W. Hodgson.

Military Cross:
Major R. Tilley.
Captain C. F. F. Moore.
Captain W. Odell.
Captain R. P. T. Ffrench.
Captain L. A. Stuart.
Lieutenant R. D. Ambrose.
Subedar Harnath Singh.

Croix de Guerre:
Lieut.-Colonel G. R. Cassels, D.S.O.

Order of the Nile (third class):
Lieut.-Colonel G. R. Cassels, D.S.O.

Companion of the Order of the Crown of Roumania:
Lieut.-Colonel G. R. Cassels, D.S.O.

Medaille D'Honneur avec Glaives en Bronze:
Subedar Major Bhura Ram, I.O.M., I.D.S.M.

Indian Order of Merit (second class):
Subedar Major Bhura Ram.
Subedar Harnath Singh, M.C.
Subedar Hans Ram.

Subedar Mohammed Alam.
Jemadar Bhim Singh.
Jemadar Arjan Ram.
Jemadar Nanig Ram.

Mentioned in Despatches :

Lieut.-Colonel G. R. Cassels, D.S.O. (2).
Lieut.-Colonel B. G. B. Kidd, D.S.O. (2).
Major R. V. Hunt.
Major A. K. Norris.
Major J. G. Rae.
Major R. Tilley.
Lieutenant C. K. Rhodes, I.A.R.O.
Lieutenant C. M. P. Durnford (2).
Lieutenant W. A. L. James.
Lieutenant W. B. Huntley.
Lieutenant H. S. Peppin, I.A.R.O.
Hon. Subedar Major Karam Dad.
Hon. Subedar Major Mardan Ali.
Hon. Subedar Dewan Ali.
Subedar Painda Khan.
Subedar Shevnarayan Ram.
Subedar Mohamed Husain.
Subedar Samrat Singh.
Subedar Partab Singh.
Jemadar Panna Ram.
2657 Colour Havildar Partab Singh.
658 Lance Naik Bega Ram.
2819 Rfm. Sheodan Singh.
2618 Rfm. Ganesh Singh.
2587 Rfm. Ganpat Singh.
2308 Rfm. Kisri Singh.
1470 Rfm. Nur Din.
776 Rfm. Hans Ram.
570 Rfm. Nanak Ram.
746 Rfm. Kalyan Singh.

Jagirs :

Subedar Major Bhura Ram, I.O.M., I.D.S.M.
Subedar Harnath Singh, M.C., I.D.M.

Indian Distinguished Service Medal:

Subedar Major Bhura Ram, I.O.M.
Subedar Samrat Singh.
Subedar Mubarik Ali.
Subedar Painda Khan.
Subedar Shevnarayan Ram.
Subedar Mohamed Husain.
Subedar Partab Singh.
180 Colour Havildar Agar Singh.
279 Colour Havildar Udmi Ram.
2657 Colour Havildar Partab Singh.
576 Havildar Ganpat Singh.
658 Havildar Baga Ram.
556 Havildar Ganga Ram.
3189 Havildar Duala Ram.
2846 Havildar Argin Ram.
224 Havildar Natha Ram.
330 Havildar Ladu Ram.
952 Havildar Mohbat Khan.
337 Havildar Nathu Singh.
608 Havildar Baz Khan.
2659 Havildar Rassa Singh.
204 Havildar Taja Ram (since killed).
226 Havildar Rup Singh.
716 Naik Sanwant Singh.
1326 Naik Ramkawar Singh.
772 Naik Ladu Singh.
738 Naik Lal Khan.
614 Naik Bhani Ram.
782 Naik Ganesh Ram.
693 Naik Lekh Ram.
2561 Naik Sujand Singh.
658 Lance Naik Bega Ram.
292 Lance Naik Gome Singh.
1504 Lance Naik Fatheh Mahomed.
799 Lance Naik Gida Ram.
2596 Lance Naik Ramchand Ram.
1071 Lance Naik Dhonkal Ram.
2153 Lance Naik Godhu Singh.
727 Lance Naik Jorawa Singh.

1315 Rfm. Shadi Ram.
2409 Rfm. Mangez Singh.
1873 Rfm. Ume Singh.
927 Rfm. Sheoji Ram.
1443 Rfm. Momray Ram.
203 Rfm. Sheochand Ram.
2414 Rfm. Ganpat Singh.
2507 Rfm. Ladhu Khan.
2803 Rfm. Tota Ram.
Syce Ramlal.
Syce Debi Din.
Bearer Kuskoo.
50 Sweeper Phulchand.

Indian Meritorious Service Medal:

2846 Colour Havildar Arjan Ram.
297 Colour Havildar Udmi Ram.
226 Colour Havildar Mukh Ram.
2960 Colour Havildar Mizza Khan.
180 Havildar Agar Singh.
556 Havildar Ganga Singh.
3189 Havildar Duala Ram.
224 Havildar Natha Ram.
330 Havildar Madhu Ram.
952 Havildar Mohbat Ram.
337 Havildar Nathu Singh.
608 Havildar Baz Khan.
2959 Havildar Rassa Singh.
294 Havildar Tejja Ram.
336 Havildar Rup Singh.
589 Havildar Kesav Rao Patoli.
2870 Havildar Algarji Singh.
92 Havildar Mehrchand Ram.
196 Havildar Jahandad Khan.
1545 Havildar Sikandar Khan.
2996 Havildar Ahmed Khan.
363 Havildar Nanu Ram.
394 Havildar Fazl Khan.
458 Havildar Hussain Khan.
673 Naik Lekh Ram.
772 Naik Ladhu Khan.

738 Naik Bal Singh.
614 Naik Bhani Ram.
782 Naik Ganesh Ram.
3132 Naik Abdul Majid.
830 Naik Johi Ram.
799 Lance Naik Gidha Ram.
2596 Lance Naik Ramchand Ram.
1076 Lance Naik Danhal Ram.
2153 Lance Naik Godhu Singh.
727 Lance Naik Jowara Singh.
938 Lance Naik Khema Ram.
292 Rfm. Gome Singh.
1504 Rfm. Fatheh Mohamed.
1315 Rfm. Shade Khan.
2409 Rfm. Mangez Singh.
1873 Rfm. Ume Singh.
746 Rfm. Kalyan Singh.
2587 Rfm. Ganpat Singh.
2910 Rfm. Raldu Singh.

APPENDIX VI

LIST OF ADJUTANTS

Names.	*Rank.*	*From*	*To*
Ogilby, R.	Lieutenant	3. 6.1820	14. 1.1822
Outram, J.	Lieutenant	15. 1.1822	19. 4.1825
Barlow, W. F.	Lieutenant	20. 4.1825	17. 8.1827
Ramsay, E. P.	Lieutenant	18. 8.1827	25.11.1827
French, P. T.	Lieutenant	26.11.1827	27. 2.1832
Brown, F. H.	Lieutenant	28. 2.1832	18. 7.1832
Cartwright, E. W.	Lieutenant	19. 7.1832	2. 7.1835
Stock, T.	Lieutenant	3. 7.1835	27.10.1844
Whitehill, S. J. K.	Lieutenant	28.10.1844	30. 8.1848
Peyton, J.	Lieutenant	31. 8.1848	10. 6.1854
Waddington, E.	Lieutenant	11. 6.1854	13. 1.1857
Anderson, H. S.	Lieutenant	14. 1.1857	30. 5.1860
Roso, W. H.	Lieutenant	31. 5.1860	5. 3.1865
Gatacre, J.	Lieutenant	6. 3.1865	20. 2.1870
Watling, J.	Lieutenant	21. 2.1870	7. 1.1879
Kellie, E. C.	Lieutenant	8. 1.1879	20. 8.1881
Scallon, R. I.	Lieutenant	21. 8.1881	24. 3.1887
Wilson, W. A. M.	Lieutenant	25. 3.1887	10. 8.1891
Delamain, W. S.	Lieutenant	11. 8.1891	22.10.1892
Dennys, A. H.	Lieutenant	23.10.1892	1.11.1894
Cumberlege, C. J.	Lieutenant	2.11.1894	4. 7.1901
Kidd, B. G. B.	Lieutenant	5. 7.1901	21. 1.1905
Hardie, G. E.	Captain	22. 1.1905	10. 1.1906
Greatwood, W.	Lieutenant	11. 1.1906	15.11.1911
Rae, J. G.	Lieutenant	16.11.1911	15.11.1915
Moore, C. F. F., M.C.	Captain	16.11.1915	18. 2.1919
Ffrench, R. P. T., M.C.,	Captain	18. 2.1919	21.11.1922
Aird-Smith, W.	Captain	22.11.1922	29. 8.1924
Greatwood, H. E.	Captain	30. 8.1924	18. 5.1928
Jones, L. B.	Captain	19. 5.1928	18. 5.1932

Appendix VII

LIST OF COMMANDANTS

Names.	*Rank.*	*From*	*To*
Hickes, J.	Major	22. 6.1820	10.1820
Gifford, F. W.	Lieut.-Colonel	10.1820	7.12.1820
Dyson, J. F.	Lieut.-Colonel	8.12.1820	8. 3.1821
Turner, W.	Lieut.-Colonel	9. 3.1821	20. 6.1822
Napier, J. P.	Captain	21. 6.1822	6. 4.1823
Bagnold, M. E.	Major	7. 4.1823	6. 6.1824
Deschamps, H. R.	Major	7. 6.1824	1. 4.1827
Scott, J.	Captain	2. 4.1827	29. 6.1827
Romkin, J.	Captain	30. 6.1827	4. 2.1833
Wilson, G. J.	Major, Lieut.-Colonel	5. 2.1833	4. 1.1840
Newport, C.	Major	5. 1.1840	4. 1.1842
Scott, J.	Major	5. 1.1842	9. 2.1842
Robertson, W. D.	Lieut.-Colonel	10. 2.1842	1. 2.1844
Scott, J.	Major	2. 2.1844	30. 4.1845
Travers, R.	Captain	1. 5.1845	30. 9.1846
Sandwith, H.	Lieut.-Colonel	1.10.1846	24. 1.1847
Travers, R.	Captain	25. 1.1847	30. 9.1847
Watkin, J.	Major	1.10.1847	10. 6.1854
Travers, R.	Captain, Major, Lieut.-Colonel	11. 6.1854	26. 4.1861
Whitehill, S. J. K.	Major, Lieut.-Colonel, Colonel	27. 4.1861	16. 1.1874
Bates, J.	Colonel	17. 1.1874	3.12.1879
Harpur, J.	Colonel	4.12.1879	3.12.1884
Gatacre, J., C.B.	Lieut.-Colonel, Colonel	4.12.1884	31.10.1891
Kellie, E. C.	Lieut.-Colonel	1.11.1891	1.11.1898
Scallon, R. I., D.S.O., C.I.E.	Lieut.-Colonel	2.11.1898	11.12.1904
Delamain, W. S.	Lieut.-Colonel	12.12.1904	12. 1.1912
Nepean, H. E. C. B.	Lieut.-Colonel	13. 1.1912	16.12.1916
Cassels, G. R., C.B., D.S.O.	Lieut.-Colonel	17.12.1916	27.12.1921

Kidd, B. G. B., D.S.O.	Lieut.-Colonel	27.12.1921	14.10.1923
Norris, A. K.	Lieut.-Colonel	1. 4.1924	31. 3.1928
Hickie, C. C.	Lieut.-Colonel	1. 4.1928	12. 7.1928
Rae, J. G., O.B.E.	Lieut.-Colonel	13. 7.1928	12. 7.1932
Ffrench, R. P. T., M.C.	Lieut.-Colonel	13. 7.1932	

INDEX

www.ingramcontent.com/pod-product-compliance
Ingram Content Group UK Ltd.
Pitfield, Milton Keynes, MK11 3LW, UK
UKHW041645190726
13854UKWH00006B/2703